Autom Inspection Systems for Industry
Scope for Intelligent Measuring

Dr. Jürgen Bretschi
Fraunhofer Institute
for Information and Data Processing (IITB)

184 Figures and 7 Tables

IFS Publications Ltd.

i

British Library Cataloguing in Publication Data

Bretschi, Jürgen
Automatic inspection systems for industry.
1. Automatic control. 2. Process control. 3. Control theory. 4. Intelligente Messsysteme zur
Automatisierung technischer Prozesse. *English*
629.8   TJ213

ISBN 0-903608-20-0

© 1981 IFS (Publications) Ltd., UK.

German edition © 1979 R. Oldenbourg Verlag, W. Germany.

Typesetting by Fotographics (Bedford) Limited, printed by Cotswold Press, Oxford, UK.
ISBN 0-903608-20-0

Contents

Preface

Measuring technology lies at the heart of automation. Traditionally most effort has been directed at developing and improving transducers and measuring procedures for the different physical variables. There are many possibilities here covered by a multitude of publications.

In recent years the lack of suitable measuring procedures and systems has increasingly placed a restraint on the development of automation. What is involved here is tasks which are reserved for humans on accont of their sensor faculties. The human faculty of vision is unsurpassed when it comes to recognition tasks such as the ordering of workpieces or visual inspection. The same applies to the assessment of noise in quality control or the use of the sense of touch in assembling and mating parts. Measuring systems for this type of task falling under the general heading of pattern recognition require a certain intelligence: apart from determining the measurement values decisions must also be made.

Intelligent measuring systems in industry represent the most recent development in the field of measuring technology. Since the beginning of the 70s more and more practical applications have been implemented. Since 1974 an impetus has been given to the development of intelligent sensor systems in the Federal Republic of Germany by the 'Humanisation of industrial work' research programme supported by the Federal Minister for Research and Technology. A part of the systems presented in Chapter 10 is a result of this research.

The motivation to write this book comes from the many years I spent as the chairman of the sensor study circle set up by the working group on handling systems. This association of industrial concerns and research institutes works within the framework of the above-mentioned research program on the development of new handling systems as technical aids for industrial processes.

This book first deals systematically with the basic principles, possibilities and limitations of intelligent sensor systems for industrial applications. No attempt has been made at completeness. In order to make this multi-discipline material more accessible the individual components of an intelligent measuring system are treated first before applying this knowledge to a number of practical examples.

The basic principles of optical, acoustic and tactile transducers are used as a starting point. A brief description is given of their operation and those characteristics of interest to industrial practice. After a short account of micro-processor systems insofar as they are important for sensor applications appropriate pattern recognition procedures are presented. This material which in the specialist literature is usually handled in a highly theoretical manner is illustrated by a number of procedures currently in use in industry and considered from a practical point of view. Much attention is paid in the last chapter to optical, acoustic and tactile sensor systems applied to industry making constant reference to the basic principles of previous chapters.

The aim of this book is to give engineers interested in measuring technology further basic information and an introduction to the practice of pattern recognition measurement data processing. The engineer entrusted with automation tasks in industry is offered help and practical stimulation in working out solutions. Last by not least the book is directed at students who are seeking to deepen their understanding in this new area of measuring technology.

Finally I would like to express my thanks to all those who have contributed to the making of this book: the Fraunhofer Institute for Information and Data Processing (IITB), in particular Dr. W. Schwerdtmann and Dr. J. P. Foith for discussions and advice and last but not least the publishers R. Oldenbourg for their kind support.

J. Bretschi

1. The concept of an intelligent measuring system

Every industrial manufacturing process is aimed at producing goods in desired quantities and with well-defined quality. In this connection measuring technology has the task of obtaining and preparing information which is necessary for optimising the production process. This industrial measuring technology, sometimes also referred to as factory measuring technology includes all measuring procedures, measuring equipment and associated activities in industrial enterprises with the exception of their development laboratories (1.1).

Industrial measuring technology can be divided into process measuring technology and production measuring technology.

Process measuring technology covers the areas of process analysis and process control. Production measuring technology concerns itself with quality control tasks and production control. Whereas process measuring technology is almost exclusively concerned with handling continuous processes this only applies in a limited manner or not at all to piece processes (discontinuous production processes) which are characteristic for production technology. Since production technology is the most important branch of industrial manufacturing technology in terms of turnover and number of employees the associated measuring technology also plays a key role.

With process measuring technology the objective is to obtain a physical variable in terms of a number and a unit. The measuring system (measuring chain) for this measurement variable usually consists of several components which are designated in Figure 1.1 in accordance with the terminology laid down in VDE/VDI 2600. This measuring system applies to all the activities or processes involved in determining by means of a comparison how many times a certain unit is contained in a particular variable (definition of measurement).

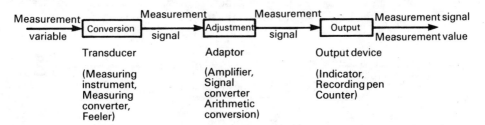

Figure 1.1   Classical measuring chain in accordance with VDE/VDI 2600.

This classical system for measuring physical variables cannot be used for a number of tasks in production. For instance the above-mentioned definition of measurement does not always apply to a checking procedure. The result of the check is no longer a numerical value in every case. As a rule the checking process is aimed at determining whether a certain property of the test object meets specific

1

requirements. This can also involve a check without using measurement, e.g. a visual, tactile or audio test. The system shown in Figure 1.1 no longer generally applies to the set of tasks with the human senses, in particular those of

- sight,
- touch,
- hearing

being particularly involved. In this connection one also speaks of optical, tactile and acoustic tasks. These tasks are most commonly encountered in production measuring technology.

A simple example where all three areas are involved at the same time is in the manipulation of a work piece. For this the position and orientation of the object must be determined visually and the hand led to the object by a tracking procedure. The tactile sense is needed for the actual gripping and finally when the object is put down the acoustic signal as well as the optical signal is processed when the object touches its support.

It is characteristic for these tasks that in contrast to the physical variables of production measuring technology the measurement variable cannot be described by a physical unit nor can the result be expressed in terms of a single number. The quantitative statement, the defining criterion of pure measurement is therefore missing. The picture of a work piece and the noise of an engine are examples of this where a complex description of the appearance or the sound impression are necessary to define the characteristics. Another feature of these tasks is that in order to arrive at the desired result both processing and discrimination are necessary which calls for 'intelligence' of the measuring system.

The Brockhaus dictionary defines intelligence as the faculty of comprehension and judgement or more generally as the complex of faculties which enables concrete or abstract problems to be solved and so allows new demands and situations to be handled. This faculty is needed when dealing with optical, acoustic or tactile recognition tasks and decision making. It therefore makes sense to talk of an intelligent measuring system. Figure 1.2 shows the general system for a measuring chain for complex recognition tasks in the optical, tactile or acoustic areas. None of the terms used in the classical measuring technology for physical variables apply.

The term sensor or sensor system will be used for such measuring systems since the tasks are analogous to those involving the human sensory faculties and these are the terms used in the English literature on the subject.

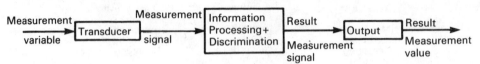

Figure 1.2   Diagram of a sensor system for recognition tasks.

2. The importance of sensors in the automation of production processes

In general the automation of a technological process means that technical means are used which allow a sequence of steps (a program) to be followed independently, i.e. without any human intervention. As a result of this program the system has the capacity to decide between different alternatives it can take in the sequence. Automation thus assumes that there is sufficiently accurate information available on the instantaneous state of the process. It can be described as an information process with acquisition of information from the environment as well as processing and outputting information to the environment. A simplified diagram of this interaction is shown in Figure 2.1. Suitable sensors obtain information from the environment and pass this to the processing stage which acts on the environment by means of effectors.

In a number of important production processes the difficulty of obtaining information on the environment is an obstacle to further automation. As already mentioned when defining the sensor concept these are especially tasks involving the highly evolved sensory faculties of sight, hearing and touch by humans.

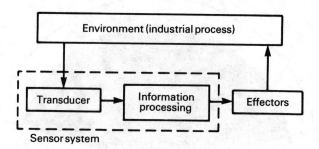

Figure 2.1   Automation of industrial processes.

There are not only economic but also social and technical reasons for replacing the sensory faculties of humans by technical sensors in the field of industrial manufacture.

Economic considerations are
– the lack of suitable personnel (an example is the specially trained test personnel for assessing sounds during quality control),
– the incomplete utilisation of labour due to periods of short employment in some cases.

Technical considerations are
– the reduction in time of the individual measurement processes,
– the removal of subjectivity in the results,

3

- improvement in quality by reducing systematic and random errors,
- easy and rapid modification,
- stricter safety regulations

Last but not least, monotonous or unhealthy work places (noise, vibrations, toxic substances, stress through excessive demands) with low social status where humans are only used for their highly developed sensory faculties need humanising (2.1). Typical examples here are conveyor belt feeding and ordering (orienting) tasks as well as the quality control of mass products.

The simplified diagram in Figure 2.1 clearly shows the key position of sensors in implementing higher levels of automation. A technological breakthrough in the field of sensors means for many production processes the possibility of increasing the extent of automation.

In particular sensors will be needed because of the increasing use of programmable manipulation devices, so-called industrial robots in automation tasks.

Up until now the automation of the work place has almost exclusively been restricted to the process itself. This meant that although the essential operating time was cut down the proportion of manipulation time in a work cycle steadily grew. Figure 2.2 shows an example of breaking down a work cycle in terms of essential operating time and non-productive time (2.2). Sensor tasks mainly occur during the manipulation phase. The relevant sectors are shown in Figure 2.2.

Humans take essential information from their environment through their eyes.

In a similar manner optical sensors in the technological field will take priority over acoustic and tactile sensors. This is reflected in the presentation of the material in this book. The resolving power of the eye is considerably greater than that of the

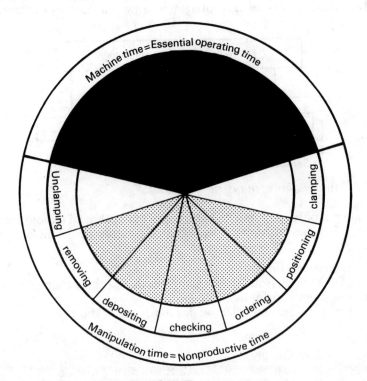

Figure 2.2   Breakdown of a work cycle (2.2).

4

hand. Two image points which are separated by one angular minute can still be distinguished by the human eye (2.3). The arc length for an angular minute is 0.0003 of the radius. At a distance of for instance 1 m the resolution of the eye is then 0.3 mm. The tactile resolving power, i.e. the spatial discrimination of two tactile stimuli is on the other hand a maximum of 2 mm at the finger tip (2.3).

3. Classification of sensor tasks

Up till now higher levels of automation have been rare in production technology in contrast to process measuring technology. In contrast to the frequently slow and continuous industrial processes a multitude of individual measurement and decisions have to be made in a short time during production.

In addition with small-scale and medium-scale production flexibility is needed i.e. changes in requirements have to be met rapidly and in an economically viable manner. This is a question of batch flexibility (fluctuations in the batch size), type flexibility (a variety of types) or technological flexibility (varying production materials).

A typical example is the ordering of parts which in large-scale manufacture is often economically viable by means of vibrating hoppers and conveyors. In small-scale and medium-scale productions the necessary flexibility when changing over to another work piece cannot be achieved in this way which means that humans have to be brought in.

The following task analysis classifies sensory tasks in groups according to increasing difficulty together with the corresponding requirements of the sensors.

Industrial production processes are characterised by a large variety of operations which can be classified as follows:

– manipulation (magazining, transporting, ordering, feeding, positioning, etc.).
– joining
– adjustment
– control (measuring, testing)
– special functions (marking, cleaning, etc.).

Just as varied are the range of tasks and requirement for industrial sensors.

Measuring tasks in production will now be classified according to increasing level of difficulty.

(a) Sensors for binary decisions
With the majority of measuring tasks purely binary yes-no statements are involved, e.g. information on the presence or absence of a part or whether a predetermined position has been reached. There is a wide range of sensors to choose from for these functions.

(b) Sensors for physical variables
A frequently occurring task in quality control is the measurement of physical variables such as geometrical dimensions, forces, pressure, temperature, etc. The level of difficulty of these tasks ranges from the measurement of a single variable to process computer-controlled measuring devices with the determination of several variables and the summary of the result in the form of a total as an intermediate level of difficulty.

One can make the general statement that as a rule there are enough accurate, high-speed and reliable sensors available. The problem is in making the checking process fully automatic.

(c) Sensors for patterns

In contrast to the above-mentioned measurements there is a number of sensory tasks which cannot be described in terms of a few numerical values. A typical example of this is the recognition of a work piece. In general terms this is a question of pattern recognition. Typical tasks of this type are:

– Determination of the orientation of a part

Example: Flexible orientation of parts when loading a machine.

Example: (1) Recognition of rejects in quality control.
(2) Positioning of industrial robots according to a special characteristic with the sensor only having to recognise the characteristic (e.g. a certain kind of hole) out of a number of other patterns present (e.g. holes with other diameters).

– Evaluation of sounds during quality control.

Example: (1) Checking the quality of ceramic tiles by tapping.
(2) Checking the operating noise of assemblies such as car engines, gearboxes, sewing machines, electric motors.

Here the sensory tasks involve reducing the large amount of information appearing in the image of the pattern to a few significant features and assigning the pattern to a particular category, e.g. good or bad.

The sensory tasks described belonging to this group have a far higher level of difficulty than the two groups considered before. One has to rely on intuitive

Fig. 3.1   Example of unordered parts in bins.

solutions to a great extent since no general theory has yet been developed.
Examples of solutions for such tasks will be considered later on.

(d) Sensors for scene analysis
A further increase in difficulty occurs when going from the above-mentioned
pattern classification to so-called scene analysis. In the literature the term scene is
used to mean a picture consisting of many objects. Whereas up till now it has been
assumed that the objects do not cover one another and are clearly distinguished
from the background this is frequently no longer the case with complex scenes in
production.

A typical example is the manipulation of an unordered workpiece in a bin. Fig.
3.1 shows the random arrangement of unordered parts in a bin. The parts cover each
other and the demarcation between the individual objects and the background is
partly missing. The image changes as the result of light reflections, shadows, dirt,
oil and rust. Despite intensive international research the technical processing of
such complex scenes is as a rule still impossible. The technical demands on
computers, memory as well as processing time stand in the way of any practical
application in production.

This book mainly deals with problems in the area of pattern recognition insofar
as they concern production. A few selected examples will be used to show how
problems in scene analysis can also be solved by means of an appropriate structur-
ing of the environment or a corresponding large investment of resources.

4. Optical transducers and image processing

4.1 Basic principles of image sensor systems

The task of the image sensor element is to convert an optical picture which is usually three-dimensional into a time sequence of electrical signals. The image sensor represents the direct link between the environment being examined and the information processing system and therefore is of particular importance. It is important for the user to know the main characteristics, the advantages and disadvantages and especially the limitations of individual sensor systems. Lack of information easily leads to simple technical applications being missed or to over-estimating the ease of application.

We shall now give a brief account of the image sensors most commonly used in industry. The emphasis is here on the properties of interest in practical applications. General principles of the individual components are only considered insofar as they are of interest for understanding variables occurring in practical use.

4.1.1. Principle of the television camera

The best known image sensor system is the television camera. A diagram of the tube of a television camera shown in Fig. 4.1. The camera lens projects an image of the scene in question through the glass bottom of the tube on to the photoconducting sensor element (target). A distinction is made between the following camera types depending on the sensor element:

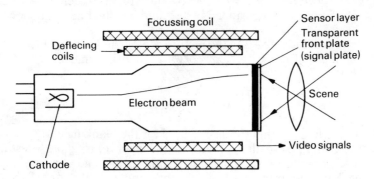

Fig. 4.1   Diagram of the pick-up tube of a television camera.

Standard Vidicon	Sensor element Sb_2S_3
Silicon diode Vidicon	Sensor element silicon (Si)
Plumbicon	Sensor element lead oxide (PbO)
Selfscanning CCD television Camera	Sensor element silicon (Si)

9

With standard Vidicon and Plumbicon the sensor layer consists of a multitude (several millions) of mosaic cells insulated from one another on a transparent metal film (signal plate). Each mosaic cell represents a small capacitor with its charge a function of the incident light. The stronger the illumination the more the capacitor is discharged.

The sensor layer is scanned with an electron beam over 625 lines in accordance with the television standard. The beam is deflected magnetically by the set of deflecting coils outside the tube bulb. The electron beam makes up the charge lost through the incidence of light in individual mosaic cells and so generates the video signal at the sensor element.

With other camera systems the capacitors in different charged states are not charged to a level but are discharged. In rare cases the video signals are not taken from the sensor element itself but from the amplitude of the return beam which is reflected from the sensor element in the direction of the cathode.

Recently silicon diode targets have gained importance as sensor layers. About 5×10^5 p-conducting islands clustered closely together are diffused into an n-conducting silicon layer about 10 μm thick (4.1). These are read out with an electron beam in a similar manner to that described above.

4.1.2 Video signal of a television camera

We shall now briefly consider the electrical video signal of a television camera insofar as it is of importance in processing industrial scenes. A detailed description can be found in (4.2).

Approximately 50 pictures per second are necessary for a flickerfree television picture. This places a heavy load on the amplifier. The same scene is therefore scanned twice. First in 1/50th of a second only the odd lines of a picture are scanned and then in the following 1/50th of a second the even lines are scanned. This means that a complete picture is established in 1/25th of a second. This procedure whereby one line is skipped in each frame is called interlaced scanning or skipping line scanning.

The number of complete pictures per second, i.e. the picture frame frequency is 25 Hz; raster field frequency is then 50 Hz. With 25 pictures each having 625 lines, $625 \times 25 = 15625$ lines will be scanned in a second, i.e. the line frequency is 15625 Hz.

The signal from a line consists of the picture content and the synchronising mark at the end of a line. With a line frequency of 15625 there is a time of

$$T = \frac{1}{f} = \frac{1}{15625} \text{ s} = 64 \cdot 10^{-6} \text{ s} = 64 \,\mu s$$

available for one line. Of this 11.5 μs are used for the blanking and synchronising signal. It consists of the so-called 'porch' and the synchronising pulse lasting about 5 μs. The synchronising pulse signals the beginning and end of a line. The porch prevents the beam jumping back to the beginning of the line flyback from appearing as a bright line. The end of a picture consisting of 625 lines is characterised by several picture pulses which are significantly different from the line pulses. During these picture pulses the picture is blanked.

The picture content is abbreviated to P, the blanking signal, i.e. the porch on the black level is abbreviated to B and the synchronising signal to S. The complete electrical television signal is therefore abbreviated to the PBS signal. The PBS signal (video signal) of a television camera is shown diagramatically in Fig. 4.2.

Between the black and the white levels the voltage of the video signal depends on the grey content of the picture.

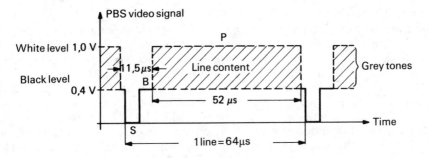

Fig. 4.2   PBS video signal of a television camera.

4.1.3   Diode line and array, selfscanning CCD television camera

Charge coupled devices (CCD) have been known as image sensors since 1970. The consequent intense research activity resulted in a large number of new technologies which eliminated the various disadvantages and problems of these new elements. In order for the user to be at least acquainted with the names of these devices they are listed below:

Charge transfer devices	CTD
Single transfer devices	STD
Bucket brigade devices	BBD
Charge coupled devices	CCD
Charge injection devices	CID
Surface charge coupled devices	SCCD
Bulk charge coupled devices	BCCD

CCD technology is most widespread. The basic principle will not be explained insofar as this applies to industrial practice (4.3, 4.4).

In their structure, charge coupled semi-conductor systems, from now on abbreviated to CCD are made up of a chain of metal oxide semi-conductor capacitors closely packed together in one plane. The basic structure is shown in Fig. 4.3.

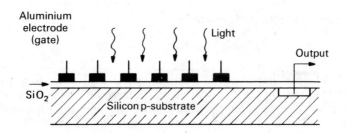

Fig. 4.3   Structure of a CCD.

11

A p-doped silicon substrate is covered with a 100-300 nm thin silicon dioxide layer above which is a row of closely packed gate electrodes. A potential well is produced by applying a voltage between the metal electrode of a capacitor and the substrate. This is where the charge carriers generated by the incident light are collected. These charges can be transferred to neighbouring electrodes in the manner of a shift register by applying suitable synchronising voltages. Several electrodes are necessary for actual charge transport: one electrode delivers the charge packet, another accepts the charge and a third prevents the charge from flowing back. With this arrangement one speaks of a three-phase CCD.

The structure of a one-dimensional or linear diode array can be seen in the diagram of Fig. 4.3. These days such diode systems contain up to 1728 picture elements. They are usually accommodated in a dual-in-line case with 22 connections. The individual diode elements have dimensions of about 16 μm × 16 μm and can be irradiated through a window. The prices range from below 100 DM to a few thousand DM depending on the number of elements, manufacturer, number of defects and batch size (1979 figures). Fig. 4.4 shows the equivalent circuit diagram for a self-scanning diode line in common use. Each element consisting of a photodiode and a capacitor connected in parallel is switched to the common video signal line by means of a MOS transistor switch. This is done by a shift register also integrated on the substrate which switches through the transistor switches one after another with the applied timing pulse. As a result of this the capacitor belonging to the photodiode in question is charged to 5 V. This capacitor charge is discharged by a photocurrent from the diode proportional to the illumination until the next scan. This missing charge is now restored during the scan controlled by the shift register. The corresponding charge pulse thereby produced on the video signal line is proportional to the illumination. The video signal of a line with N elements is therefore a sequence of N pulses with amplitude proportional to the illumination.

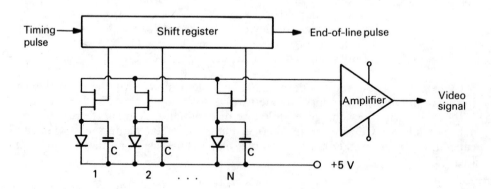

Fig. 4.4   Circuit diagram of a diode line.

In most practical applications a single operational amplifier is sufficient to amplify the video signal. It is operated as a charge or current amplifier. Fig. 4.5 shows the pulse-shaped video signal of a diode line with the illumination blocked by an object to illustrate the change in level in the middle section.

At the end of the shift register (Fig. 4.4) an end-of-line pulse is generated to signify the completion of a line.

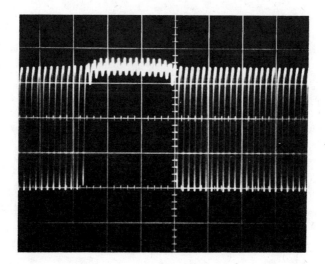

Fig. 4.5   Video signal of a diode line (example).

The arrangement of such diode lines over a surface results in two-dimensional diode arrays. There are various possibilities of reading out the information from such systems.

With the originally used xy read-out the elements are assigned to vertical and horizontal tracks and read out. The principle is shown in Fig. 4.6.

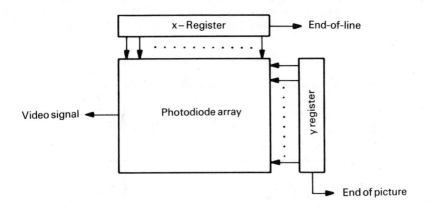

Fig. 4.6   Read-out of a photodiode array.

An end-of-line pulse is generated at the end of the x register, available as a synchronising signal. At the same time a counter for the y register which can be connected externally is incremented and the x-register run through again. At the end of a picture, i.e. at the last position of the x and y registers the end-of-picture pulse appears at the output of the y register. Fig. 4.7 shows a simplified block diagram for the generation of a video signal from a diode array. The manufacturers of diode arrays offer ready kits for these circuits.

13

A basic defect of this read-out mode is the limited number of elements because of the large capacity of the read-out line. This is why this read-out mode is only found with arrays with a small number of elements, e.g. 50×50 elements.

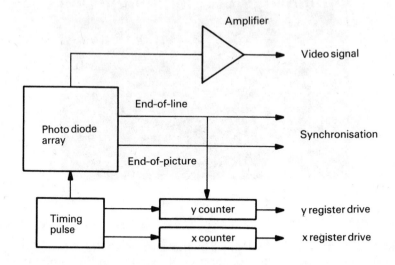

Fig. 4.7   Block diagram of a diode array drive.

A better procedure is to first store the picture information in a buffer and only then to output it. This is shown diagrammatically in Fig. 4.8 for a 512×320 CCD array. The charge pattern corresponding to the optical picture is very rapidly transmitted in parallel to the buffer which is connected to a special CCD line, the output register.

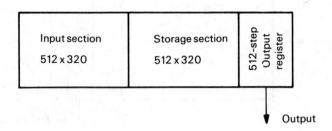

Fig. 4.8   Diagrammatic representation of a CCD arrangement with 512×320 elements.

It is the aim of CCD development to produce arrays with the necessary resolution for a television picture which would mean approximately 600×600 picture elements. There is no doubt that this will be achieved in a few years.

The advantages of such a camera are obvious. The picture is free of all the defects of electron beam scanning such as geometrical faults, mechanical sensitivity, short

14

working life, large dimensions, large weight and high power consumption.

The 512 × 320 step CCD arrangement (RCA SID 52501) is the largest array available at the present time (1979). Each of the 163840 sensor elements has a size of 0.03 mm × 0.03 mm and they are accommodated on a 19 mm × 12.5 mm semiconductor chip.

A brief description will now be given of the functioning of a CCD television camera using this read-out mode.

At the end of a 1/25 s picture duration the picture is simultaneously transmitted to the buffer during the blanking of the frame repetition pulse. The first line is transmitted to the output register at the end of the blanking (Fig. 4.8). This register is read out inside 52μs (Fig. 4.2) corresponding to the duration of a television line. Then the next lines are successively read out via the register until all the lines of the stored picture have undergone this process. According to the television standard there are 11.5 μs (duration of the horizontal blanking pulse) for inputting each line into the output register.

Charge-coupled image sensors of this type, especially television cameras will in the future be highly competitive with conventional camera tubes.

The costs which at present are still high will be so low in a few years that this camera type will become extremely widespread because of its advantages.

Two-dimensional arrays with more than 100 × 100 elements are at present (1979) not yet suitable for many industrial applications because of the high price and many defects which show up as breaks in the picture. Television cameras with electron beam scanning are therefore still much superior when it comes to tasks requiring resolution.

4.2 Image sensors as measurement devices

Image sensor systems serve in the first instance to observe processes in places separated at a distance. The best known example is the home television set.

However in the same way it is possible to carry out a number of measurements with respect to the object using the information contained in the video signal.

When a scene is scanned over two dimensions one picture point after another is converted into an electrical signal with an amplitude proportional to the brightness of the scanned point. Thus it is possible to determine the light intensity of any picture point as a first measurement variable.

Another measurement variable which can be determined is the position of an optically prominent point. With line scanning of the picture with a television camera this position can be defined for preset picture and line frequency by the point in time when a predefined brightness step occurs in video signal. This is illustrated in Fig. 4.9. The top part shows the scanning of a point to be measured. The sawtooth deflection voltages U_x and U_y are shown in the middle. These guide the electron beam over the picture. Due to the fixed relationship between the temporal sequence of the scanning and the position of the electron beam the amplitude of the two deflection voltages also defines the position of the picture point. The voltage amplitudes at the point in time of the pulse from the picture point in the video signal (Fig. 4.9c) are proportional to the position of the picture point.

With a diode array or line the position of the point is given directly by the fixed relationship of the diode elements to the corresponding picture points.

Any picture point with a characteristic intensity can serve as an optically prominent point, since with the help of electronic thresholds all signals with

15

amplitudes in a certain range, i.e. all picture points with a certain brightness can be sorted. An example of this is the procedure discussed in section 4.5.1 for generating a binary picture.

In a modified procedure the lines inside a picture are continuously added and externally generated timing pulses counted from the beginning of each line to the picture point. The state of the counts are then directly proportional to the position of the picture point.

If several picture points are measured then section lengths can be measured as another measurement variable.

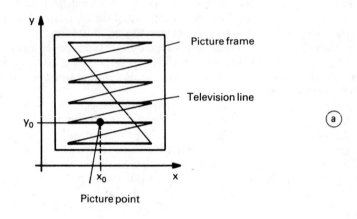

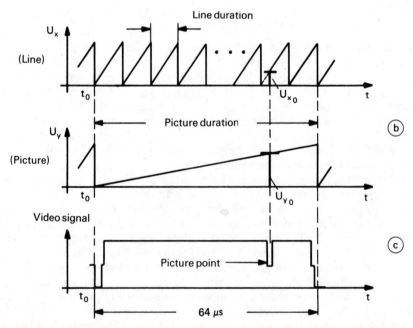

Fig. 4.9   *Measurement of the position of a picture point with a television camera.*

(a) Picture scanning
(b) Deflection voltages for line and picture
(c) Video signal

16

If for instance two picture points occur in a line the distance between them can be determined from the difference between the U_x deflection voltages or the number of measured counting pulses between the two picture points.

If the distance between the individual lines is known then the area of an object can also be determined from the section lengths in the individual lines with the help of the state of the line counter.

In addition by using suitable filters it is also possible to apply all the above-mentioned measurements selectively to certain colour components of an object.

4.3 Characteristics of properties of image sensors

The most important properties of the image sensors covered in the previous section will now be examined with respect to their industrial application. In particular the following characteristics will be considered in detail.

- Resolution (definition)
- Spatial modulation transfer curve
- Geometrical faults
- Sensitivity
- Transfer linearity
- Lag
- Inertia
- Spectral sensitivity
- Blooming
- Noise
- Long-term stability
- Scanning time
- Mechanical and electrical data

The classification of these characteristics makes it possible for the reader to understand the concepts appearing in the various data sheets and to assess their importance. At the same time the quantitative data which are scattered throughout the literature serve as a useful guideline.

4.3.1 Resolution and spatial modulation transfer curve

The resolution of an image sensor can be defined as the number of picture elements which can be discriminated. The resolution of a television picture is essentially limited by the number of lines and the frequency band. The television picture consists of 625 lines. The ratio of the width of the picture to the height is 4:3. If the picture points to be transmitted have the same dimensions in both directions then 625.4/3 = 833 picture points are covered by one line width. This gives 625·833 ≈ 520,000 picture points for the whole picture. Since 25 pictures are transmitted in one second (picture frequency) this gives a rate of approximately $13·10^6$ picture points per second. If these picture points are alternatively black and white then a picture voltage is produced of $13·10^6/2$ = 6.5 MHz since a bright and dark picture point represent a cycle. Since this extreme case is not permanently present in practice the highest frequency of the video signal to be transmitted is restricted to 5 MHz which corresponds to about 10^7 picture points per second or – for 25 pictures per second – to 400,000 picture points per picture. With 625 lines this gives in the most favourable case about 640 picture points per line which can be resolved.

Roughly speaking the resolution limit of a television line for practical applications is 8 bits. As a rule of thumb one can say that the resolution of a standard industrial television camera amounts to about 1% of the picture area.

A common method of characterising resolution losses is to specify the depth of modulation which is present in the signal when transmitting a vertical running black-white stripe sequence. The variation of the modulation depth as a function of the number of stripes per unit of length, the so-called spatial frequency is characterised by the spatial modulation transfer curve (4.5). Frequency in the time domain can be defined as the number of oscillations per time unit. If this is transferred to the spatial domain of a picture then the spatial frequency can be defined as the change in picture intensity over a certain unit of length. In this case as already mentioned each sequence of a dark and white picture point represents a cycle.

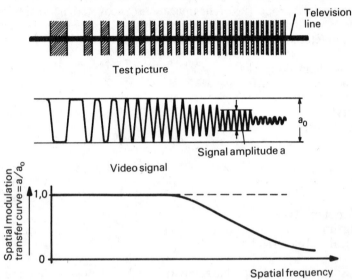

Fig. 4.10   Illustration of the resolution by the spatial modulation transfer curve.

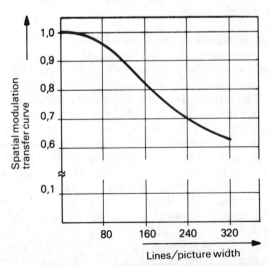

Fig. 4.11   Typical modulation transfer curve of a television camera.

18

The definition and main features of the spatial modulation transfer curve can be taken from Fig. 4.10. A typical modulation transfer curve which more or less applies to all types of camera is shown in Fig. 4.11.

4.3.2 Geometrical faults

For television cameras with electron beam scanning deviations in the constancy of vertical and horizontal deflection show up as faults in the geometrical assignment of the picture content. Typical geometrical distortions are linearity defects in the vertical deflection which extend the picture upwards like a distorting mirror. Similarly a widening of the picture takes place with faulty horizontal deflection.

Standard industrial cameras are not designed as measuring cameras but in order to generate a picture for examination. They therefore exhibit relatively large geometrical faults. For a standard industrial television camera this is usually ± 1% − ± 2% (with respect to the picture frame). With cheap cameras this fault can easily be larger.

With CCD television cameras there are no geometrical faults since there is no electron beam scanning.

4.3.3 Sensitivity and transfer linearity

The input signal of an image sensor is a brightness distribution. The output signal is a current or voltage proportional to this brightness. The sensitivity is defined as the ratio of the output magnitude to the input magnitude (dimension e.g. μA/lumen). Quite generally the following applies (4.6).

$$\text{Output magnitude} = (\text{Input magnitude})^{\gamma} \qquad (4.1)$$

where gamma the exponent of the transfer function is given by

$$= \frac{\log(\text{output magnitude})}{\log(\text{input magnitude})} \qquad (4.2)$$

For a linear detector $\gamma = 1$. The exponent for different image sensor systems is specified in the following table together with the sensitivity. These are typical values (4.6, 4.7).

As a rule the standard Sb_2S_3 Vidicon has roughly the same sensitivity as the Plumbicon. The semi-conductor cameras are however far superior to both camera types in terms of sensitivity.

Image sensor	Sensitivity μA/lm	Gamma
Sb_2S_3 Vidicon	40 – 1200	0.6
PbO Plumbicon	300 – 400	0.95
Silicon Vidicon	4000	1
CCD camera	3500	1

Table 4.1   Sensitivity and γ for image sensor systems.

4.3.4. Lag

Lag is defined as the percentage of the signal current at a certain point of the target after the illumination has been switched off. Typical values for the lag are shown in Table 4.2 (4.6, 4.8).

Image sensor	Lag
Sb_2S_3 Vidicon	20%
PbO Plumbicon	2%-5%
Silicon Vidicon	8%
CCD camera	1%

Table 4.2   Lag for image sensors

Fig. 4.12 shows the lag for a plumbicon and a standard vidicon for the case of strong illumination. The Vidicon has a considerably larger lag than the Plumbicon.

The lag means a limitation in the practical industrial application of television cameras in terms of the permissible speed of movement of an object under consideration. This case is most frequently encountered with parts being transported on a conveyor belt.

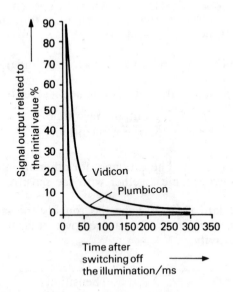

Fig. 4.12   Lag for Plumbicon and standard Vidicon (4.9).

If in the simplest case the monitor shows up an impermissible movement blur due to the lag effect it is possible to illuminate the object briefly with a flash. In this way the lag is specifically exploited in order to store the picture after the flash for the duration of the picture processing. In the same way it is possible to carry out the short-term scanning of the object by means of a rotating aperture in front of the camera lens.

In departure from the electron beam scanned television cameras discussed up till now the problem of movement blur is absent if a sensor line is used for the moving object. With the help of the sensor line the object is very rapidly scanned in the coordinate vertical to the movement with a frequency which ranges from 100 kHz to a few MHz. Because of the movement of the object itself under the sensor line there is also no need for spatial scanning by the sensor. Thus a one-dimensional line is sufficient to process a two-dimensional picture.

4.3.5 Inertia

The inertia like the picture definition can be determined by means of a function, namely the temporal modulation transfer curve. This shows how a sinusoidal illumination of the target varying over time is transferred as a function of its time frequency. With an inert system a rapid time change at the input of the system appears at the output with reduced amplitude. Fig. 4.13 shows the temporal modulation transfer for a $Sb_2 S_3$ Vidicon and a Pb_0 Plumbicon (4.10).

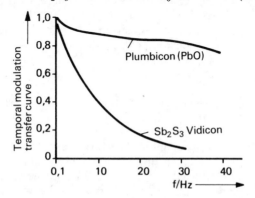

Fig. 4.13   Temporal modulation transfer curve for image sensors (4.10).

4.3.6 Spectral sensitivity

The spectral sensitivity of an image sensor system is defined as the variation of the output as a function of the wavelength of the incident light. Normally the spectral sensitivity is of special interest in the visible region. However it can be just as important to give preference to a spectral region outside the visible region. An

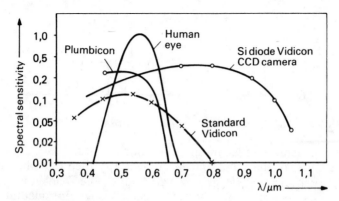

Fig. 4.14   Spectral sensitivity of different image sensors in comparison to the human eye.

21

example is the observation of red-hot parts when especially high sensitivity in the near infra-red is an advantage. Fig. 4.14 shows the spectral sensitivity for different image sensor systems as well as for the human eye.

It can be seen that the semiconductor cameras are distinguished by good sensitivity in the invisible near infra-red region.

4.3.7 Blooming

If the target of a television camera is subjected to intensive brightness then the excess charge carriers spread into neighbouring zones and introduce the bright area there. This effect is called blooming. No figures are given in the literature or in manufacturers' documentation.

It is important for the user to watch out that bright points on an observation monitor do not merge and appear wider than the actual geometrical image.

It is especially important that the target is not destroyed by overexposure. Semiconductor cameras have distinct advantages here. Whereas with standard Vidicon and Plumbicon a local overexposure destroys the affected point of the target semiconductor cameras have extreme resistance to exposure so that they should always be used where strong local exposure effects are expected. An example of such an application is given in section 10.1.10.

4.3.8 Noise

The statistical fluctuations accompanying the physical process of picture generation show up in the television picture as a granular background which is called noise. Noise is defined quantitatively by the signal-to-noise ratio (S/N), i.e. the ratio of the amplitude of the wanted signal to the average amplitude of the interference. For television cameras the signal-to-noise ratio is defined as the peak-to-peak video signal divided by the effective value of the noise (4.11). The video signal of a noisy television line is shown diagramatically in Fig. 4.15. The amplitude of the wanted signal is referred to the midpoint of the noise signal amplitudes and is determined from the distance CD. The effective value of the noise is 1/6th of the noise amplitude AB (4.11). Thus

$$\frac{S}{N} = 6 \cdot \frac{CD}{AB}$$

$$(4.3)$$

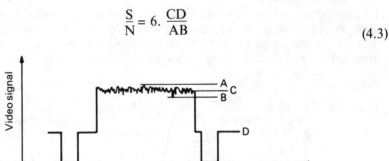

Fig. 4.15   Determination of the signal-to-noise ratio.

If one follows general practice in telecommunications and expresses the ratio of electrical variables of the same unit logarithmically then one obtains the level. Level is expressed in Decibels (abbreviated to dB). The Decibel is defined as 20 times the decimal logarithm of a ratio of linear variables (for quadratic variables, e.g. power, 10 times the decimal logarithm is used). The signal-to-noise ratio of the video signal is also expressed in dB. The following reference points can serve as

practical guidelines. A signal-to-noise ratio of 20 dB means that the pictu
is very bad – there is inadequate detail and resolution. For satisfacto
quality S/N must be 40 dB and more. The limit for noise detection lies ;
dB.

4.3.9   Long-time stability

The stability of the electron beam system determines the stability of the scene scanning by the television camera. A possible measure for the stability of scanning is the temporal geometrical deviation of a line from an initial position. Studies have shown that for an industrial television camera in the middle price range the line deviation is about ± 1% of the picture area over a long period of time.

This fault is impossible with the self-scanning CCD cameras.

In this connection it is interesting to look at the behaviour when switching on. With industrial television cameras in the middle price range the final stability of the lines is reached after a maximum of 1 minute. There is no such switching-on delay for CCD television cameras.

4.3.10   Scanning time

The picture frequency of a television camera with electron beam scanning is 25 Hz in accordance with the television standard (cf. section 4.1.2). Therefore a time of 40 ms is needed to scan an entire television picture. Assuming a speed of 25 cm/s a moving workpiece on a conveyor belt has changed its position by about 1 cm at the end of a picture scan. This value is further increased by the picture processing time. If the conveyor belt is constant then this spatial displacement can be taken into account by suitable calibration.

This simple numerical example shows that with electronic picture scanning by means of a television camera delays are produced due to the serial scanning of every line and these must be taken into account with moving objects. Diode lines and arrays allow for a faster scanning of the picture depending on the timing pulse applied.

4.3.11   Mechanical and electrical data

The weight of a standard industrial television camera is about 4 kg; its power consumption is about 25 W. The weight of the smallest electron beam scanning camera is about 1 kg and the power consumption about 4.5 W. The CCD cameras which in principle can be made very small also fall into this range at the present time.

The weight together with the corresponding geometrical dimensions is too large for applications in which for example the camera has to be incorporated in the gripper of a manipulation system. A solution for such a case is to transmit the picture information to the television camera by means of optical fibres.

4.4   Evaluation of image sensor systems in practice

In principle all of the above-mentioned image sensor systems are suitable for industrial use. As a rule one should always use a standard Vidicon wherever possible because of the price. Only where considerable overexposure is expected should a Si diode Vidicon which is twice as expensive be employed.

The use of a Plumbicon in industry brings no advantages compared with a standard Vidicon. The Plumbicon is mainly in television studios or as a colour television camera.

CCD cameras are still very expensive (twice as expensive as the Si diode

Vidicon). Their advantages however are so evident that their place in future is assured.

The use of diode lines and arrays is especially indicated when there are problems of space and weight. Because of picture distorting defects the number of elements is at present restricted to 100×100 for practical applications. However there is much activity in this very area of technology so that corresponding improvements can also be expected here.

4.5 Methods of information reduction (contraction) during picture processing.

Based on the diagram shown in Fig. 1.2 of a sensor system for recognition tasks the sensor generates a signal and transfers it to a stage which processes this information. It stands to reason that as a rule the more data supplied by the sensor the more complex is the information processing.

Because of the high speeds of machines in production real-time processing is necessary which is only achieved if there is as little data as possible coming from the processing and discrimination stage. Especially with data-intensive picture processing it is therefore necessary to reduce the information either already before taking the picture or during this process. Some of the methods of information reduction for picture processing will now be presented (4.12).

4.5.1 Illumination and type of picture

The following types of pictures can be employed for picture processing:

- colour pictures
- grey-tone pictures
- binary pictures

The natural surroundings of humans presents itself in the form of a colour picture. Although the possible range of colours in industry is fairly restricted its processing represents a very high level of difficulty. The processing stage of the sensor has to process three grey-tone pictures which correspond to the colour separations red, green and blue. These three pictures have to be processed simultaneously for reasons of time. Up until now the processing of colour pictures has not been carried out because of the high level of difficulty.

Grey-tone pictures exhibit 32-256 grey-tones. Reference can be made here to Fig. 3.1 as a typical example of a grey-tone scene. It can be seen that the separation of the individual objects (segmentation) is very difficult. In practice there are further complicating problems caused by disturbances such as dirt, rust, light reflections, etc. Therefore grey-tone pictures are only rarely processed in industry at the present time.

It is much simpler to use binary pictures. They only exhibit the brightness steps 'black' and 'white'.

If these pictures are reduced to the edges of objects where there is a change in brightness value one obtains a contour picture of the object.

Fig. 4.16 shows the grey-tone picture of a workpiece and the corresponding binary and contour pictures.

In order to generate a binary picture the object must be clearly separated from the background by appropriate illumination. There are two possibilities of doing this:

(a) transmitted light
(b) direct light

An example of the transmitted light method is a transparent conveyor belt illuminated from underneath, transporting the workpieces past the sensor (Fig. 4.17). An example of this will be considered in section 10.1.5.

With direct light a distinction is made between diffuse and directed illumination (Fig. 4.17). The diffuse incident light corresponds to daylight illumination and is achieved industrially with large-area light sources (cf. the example of section 10.3.6).

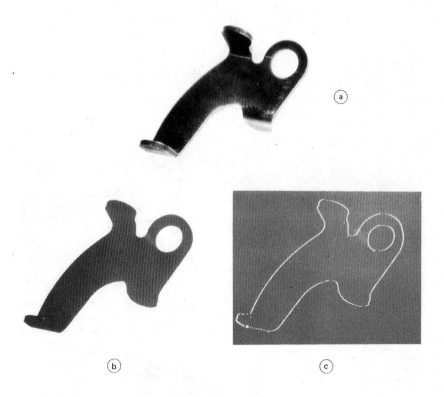

Fig. 4.16 Grey-tone picture (a), binary picture (b) and contour picture (c) of a workpiece.

If with directed light one looks exactly in the direction given by the law of reflection (blazing angle) then even surfaces with different colours come out bright in the same way. Even black parts appear bright (bright field observation). Interruptions to the surface, e.g. holes in a plate do not relect the light and therefore appear as dark spots in the picture.

With dark field observation the object is observed at an angle which is quite different from the blazing angle. Disturbances in the surface can be recognised here from the diffused light.

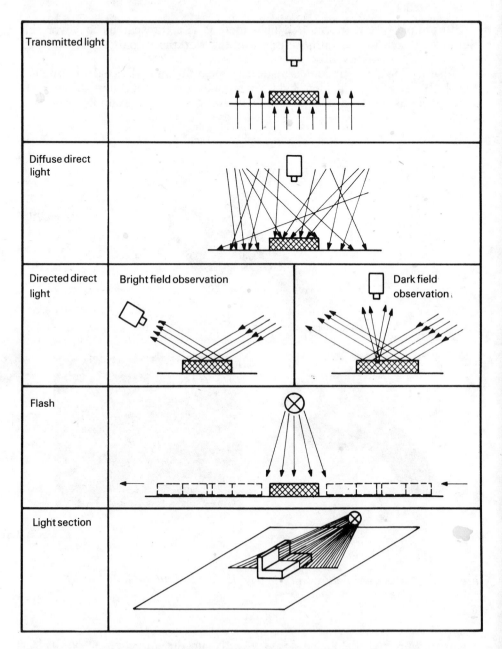

Transmitted light	
Diffuse direct light	
Directed direct light	Bright field observation / Dark field observation
Flash	
Light section	

Fig. 4.17   Problem-oriented illumination methods (4.13).

What type of illumination is used depends on the specific task and often cannot be determined from the outset. The simplest way of deciding which illumination is

26

best is to make preliminary tests. The picture to be processed is taken for instance with a television camera and the resulting images evaluated on a control monitor on the basis of the recognisability of desired details.

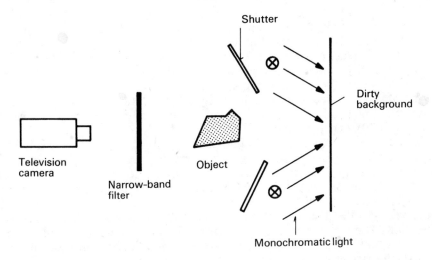

Fig. 4.18 Generation of a binary picture by means of monochromatic illumination.

As a rule in industrial practice the background of the object for optical processing cannot be changed arbitrarily. For instance a workpiece clamped in a lathe is badly distinguished from the background of the machine and is often heavily soiled by drilling liquid, swarf, etc.

One possibility of improving object recognition is to illuminate the background with monochromatic light (e.g. a sodium vapour lamp) (Fig. 4.18). A narrow-band filter placed in front of the television camera and matched to the wavelength of the light source makes the background appear light and the object appear dark.

Another possibility of simplifying the scene is to generate a light-section picture (4.14). These pictures are a special case of binary pictures. They are slit images of the observed scene and are formed by projecting a gap or a slit on the scene (Fig. 4.17). The projection of the slit adapts itself to the structure of the scene so that the course of the projection reflects information on the three-dimensional properties of

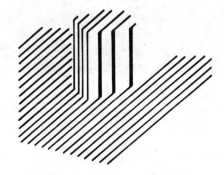

Fig. 4.19 Light-section picture of a cuboid (4.15).

the scene (cf. Fig. 4.19 for a light-section picture of a simple cuboid). In order to obtain information on the entire scene the slit is either moved sequentially over the scene (4.15) or a series of slits imaged in parallel (4.16). With light-section pictures there is no dependence on the variation in brightness and information is obtained on the scene directly.

When processing light-section pictures the projection lines must be followed in order to determine the points of inflection. These can be used to deduce the geometrical coordinates of the scene. There are problems when the projections widen out or contract, are interrupted and run into each other. The correct assignment of the lines to one another is of key importance.

The information reduction produced by a light-section picture can be clearly seen in the practically important case of the automatic monitoring of a welding path (Fig. 4.20). In this example the edge of the bracket attached to a plate is to be processed as the path of the weld. Light-section pictures are especially suitable for processing such edges and following them since they yield the 3D structure of the scene immediately. Since in this case the position of the welding path can already be deduced from the projection of a single slit there is no problem of assigning the slit projections to one another. The projection of a slit perpendicular to the welding path gives a profile which corresponds to a cross-section through the parts along the projection. The welding path appears in profile as an indentation, inflection or irregularity or interruption. Instead of a complex grey-tone picture one only has to evaluate the path of a set of curves.

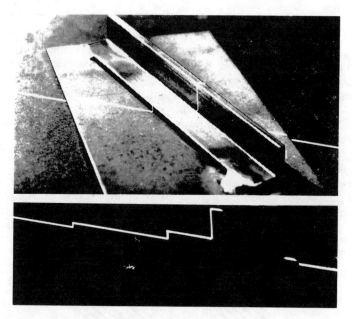

Fig. 4.20   Information reduction by slit illumination (4.17).
            (a) Grey-tone picture with slit light.
            (b) Path of the slit light.

The problems of widening, narrowing or disappearance of the projections can be avoided to a large extent by projecting a surface instead of a slit so that one half of the scene is bright and the other half lies in the dark.

Profiles formed by the projection of a surface are similar to those formed by the projection of a slit but there are no interruptions or widening out. The projection of a surface has the additional advantage that a binary picture is produced in which the boundary between light and dark represents the profile to be evaluated. Fig. 4.21 shows the corresponding binary picture for the example from Fig. 4.20.

It cannot always be guaranteed that a single boundary will be produced; however as a rule branching can be logically detected and eliminated. If one now regards the boundary between light and dark as a contour line the problem arises of analysing the shape of a contour. In this way the analysis of light-section pictures can be naturally incorporated in the evaluation of binary pictures.

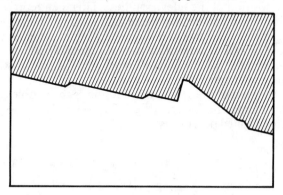

Fig. 4.21   Light-section picture as a binary picture (4.14).

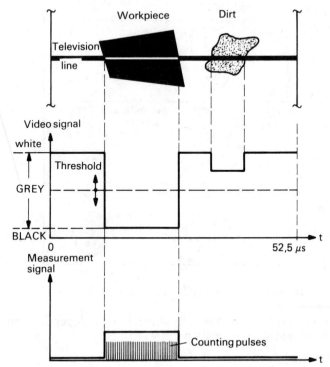

Fig. 4.22   Generation of a binary picture by a threshold value operation and signal processing.

29

For the sake of completeness the flash illumination of moving objects considered in section 4.3.4 is also included in Fig. 4.17. By taking an instantaneous picture a stationary image of the object is produced which can be used for the processing.

4.5.2   Binary picture evaluation

A picture can be often converted to a binary picture by a simple electronic threshold operation. Fig. 4.22 shows the video signal of a television line for a predefined pattern. The dirt lying below the threshold value is ignored for subsequent processing. The number of counting pulses can serve to characterise the workpiece at the position of the television line. This then represents a measure of the geometrical dimension of the workpiece at the scanned point (section 4.2).

If the extent of soiling is so large that the corresponding video signal overlaps that of the workpiece it is no longer possible to make clear separation. This separation must then be made on the basis of suitable workpiece characteristics and by means of special procedures (Chapter 8).

An important method of information reduction is the simple description of the picture content using specific features (Fig. 4.23). A good example is the number of counting pulses mentioned previously.

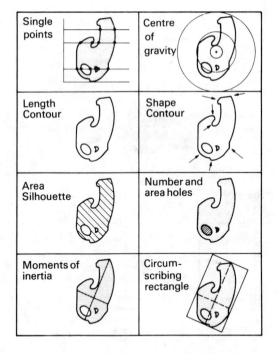

Fig. 4.23   Possible methods for describing pictures (4.14).

These features can be divided into three groups depending on how much information is evaluated from binary picture (4.14):

- single points
- contour lines
- silhouettes

If only single points are analysed these must be referred to a reference system. This can either be object-bound or object-free.

With an object-free reference system the position of the object with respect to the reference must be specified in a defined manner (cf. the example of section 10.1.5).

An object-bound reference point must exhibit certain invariance properties. Such a point is the centre of gravity which in addition relatively insensitive to disturbances to the contour since it is calculated as an integral. The position of the centre of gravity can be used as a feature for recognising position.

If circles are made about the centre of gravity then the points of inter-section of the circles with the silhouette can serve as features. A sensor system which works with such features is considered in section 10.1.6.

Other features which are used are: the radius of the largest circle which still just touches the contour; the average radius from the centre of gravity to all points of the contour.

The length and the shape can be evaluated from the contour line. Whereas the length serves as a feature for the classification of different workpieces, the shape can be used to deduce the position. However there are only a few procedures which can be used to evaluate the shape. The shape of the contour can be described by indentations, inflections, projections, etc.

The area and shape can be evaluated from the silhouette. As a rule the shape is difficult to process. The shape of a silhouette can be stored as a mask and then an unknown silhouette can be compared with different masks (Section 8.2.2.2).

Analytical relationships are given in Section 8.2.2.1 for some important features.

4.5.3 Mechanical aids
According to the above considerations it is very difficult and time-consuming to process a complex grey-tone picture which is produced for example when parts lie disordered in a box. However very often in industry it is possible to structure the scene so that it can be readily imaged.

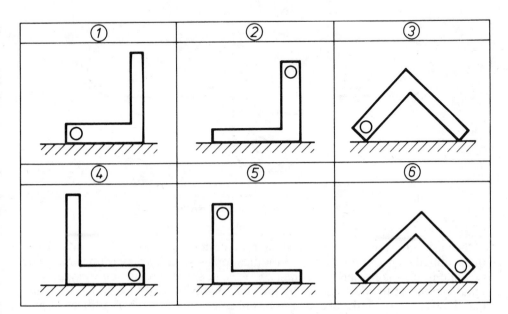

Fig. 4.24   Stable orientations of a workpiece at a mechanical stop.

The task of 'reaching into a box with disordered parts' becomes considerably simpler if for instance it is possible to isolate the parts and reduce their number of degrees of freedom.

In industrial applications it is often possible to isolate parts by means of mechanical appliances. The sensor then only has a single workpiece to process.

In addition by making a further restriction in the number of degrees of freedom the number of possible workpiece orientations can be reduced. The simplest restriction of the number of degrees of freedom is to deposit the parts on a flat surface (table, conveyor belt). If in addition the workpieces are allowed to run against a mechanical stop then a few stable orientations of the workpiece are produced as shown diagrammatically in Fig. 4.24 (4.18). The picture of the workpiece reduced to a few possible orientations is considerably simpler to process than the infinite variety of disordered parts. Wherever possible picture processing should be facilitated by taking these kind of simple measures involving mechanical aids.

4.5.4 Recognition-orientated workpiece design

In many practical cases the sensor task is considerably simplified by suitable design of the workpiece (4.19). The recognition of objects depends entirely on how well the object features can be discriminated. Often the binary picture does not distinguish between different resting positions despite considerable structural differences within the workpiece. The workpiece shown in Fig. 4.25 is completely symmetrical and so if there are structural differences on the surface such as small protuberances, indentations, burr on the edges, roughness, grain or stains on only one side in each

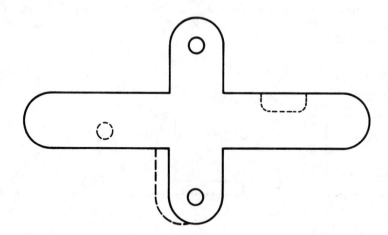

Fig. 4.25   Example of a recognition-orientated workpiece design.

of the positions cannot be distinguished from the contour picture. A disturbance of the symmetry for example by the modifications shown by dotted lines makes it possible to distinguish between the two sides. A specific reduction in symmetry made already at the design stage by introducing functionally meaningless details entails no extra cost and makes an important contribution to the simplification and feasibility of the sensor task.

4.5.5 Sensor programmed learning

Very roughly speaking a recognition process can be described as the comparison of a presented pattern with a known pattern which has been stored in memory. Thus the learning of reference patterns is a prerequisite for recognition.

With sensor imaging the possibility of easy learning of object-specific features is a condition for the application of the system to meet changing requirements.

Learning can be defined as the acquisition or improvement of a property or ability of a system in interaction (information exchange) with the environment. This can take place under the direction of a teacher using an objective or success criterion (4.20).

From the definition of learning it follows that on the one hand there is an interaction of the learning system with the environment and on the other hand there is positive or negative reinforcement of the system by the teacher or success criterion.

Normally the human teacher only appears during the development and design of measurement systems and a number of industrial recognition tasks. The measurement task is clearly defined – its execution is exactly tailored to the particular application so that the system does not learn.

However especially in production there are many tasks where the recognition system must adapt itself very rapidly and at low cost (i.e. with great flexibility) to changing requirements (e.g. type changes).

In this case it would be too cumbersome to cater for all possibilities in the sensor in advance. It makes more sense here to prime the sensor with a basic structure and to adapt it to new objects by a simple learning process. Based on the system shown in Fig. 1.2 for recognition tasks the learning process can be followed using Fig. 4.26.

Acting as a teacher the user fixes the object-specific selection of features. This feature selection as well as the comparison model derived from it is variable and is redefined for each object. Based on the result the user may find it necessary to modify the feature selection and so correct the comparison model to obtain a more accurate and reliable recognition. The comparison model for optical imaging

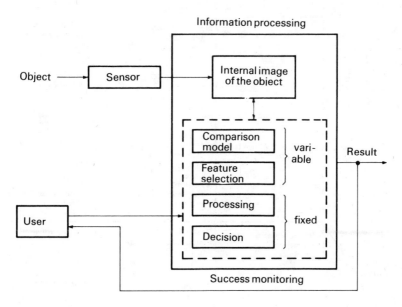

Fig. 4.26   Learning process of an image sensor.

systems can be put together by 'presenting' all the possible views of the object. The picture taken by the sensor is stored in memory in term of the feature selection.

During the actual activity phase, also called the 'can' phase the presented objects are compared with the previously learned model and the result obtained from this. This method of programming (teach-in programming) considerably simplifies sensor processing. The programming is carried out with the help of the *a priori* knowledge of the object to be processed. Examples of this kind of sensor programming are given in sections 10.1.4, 10.1.5 and 10.1.6.

The next step is learning without a teacher with the feature selection made autonomously and optimally in the sense of recognition. However the implementation of such self-learning systems is costly and is at present not used in industry.

4.6 Limits of industrial picture processing

The previous considerations show that the level of difficulty of picture processing depends essentially on three parameters.

    – the number of grey tones
    – the number of mutual disposition of the objects in the picture field
    – the number of degree of freedom of the object

The function of picture processing is to segment the picture into isolated objects, classify the objects and determine the orientation class and position parameters. Often there is no problem of segmentation and classification with only one workpiece in the picture so that only its orientation and position need be determined.

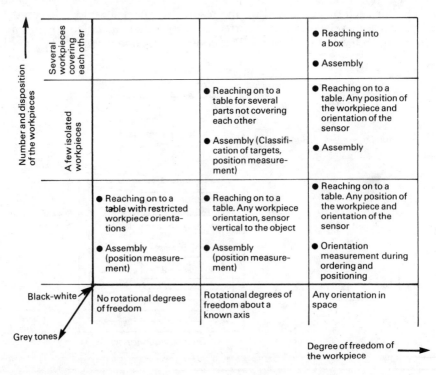

Fig. 4.27 *Level of difficulty of industrial recognition tasks (4.12).*

There are additional classification problems with several isolated workpieces since the different workpieces have to be distinguished. If workpieces touch each other or overlap the problems of segmentation and classification become extremely difficult. A variety of disturbances makes it difficult to find separating lines between the workpieces. An added difficulty in classification is that the workpieces are often only partly visible.

The number of degrees of freedom of the object determines how many orientation classes should be distinguished. The processing effort increases with the number of orientation classes and position parameters. With only a few orientation classes it is possible by template comparison to distinguish between the classes. With increasing numbers such procedures become unwieldy or fail completely.

Fig. 4.27 lists sensor tasks with varying levels of difficulty in terms of three key parameters. Some typical industrial tasks are given by way of illustration. In general the task becomes more feasible if it is possible by means of suitable technical measures to limit the degree of freedom of the objects, to prevent mutual obscuring, to isolate parts and to reduce grey-tones to a binary picture. This means with reference to Fig. 4.27 that the level of difficulty of the task decreases as one approaches the centre of the coordinate system defined by the three parameters.

Many recognition tasks which are very simple for humans can still not be accomplished with technical picture sensors. The implementation of sensors with the full capacity of the human visual system is not possible scientifically or technically in the foreseeable future. However by exploiting specific conditions of the task in question and systematically applying information-reduction methods as well as dividing the task up into several feasible subproblems technical solutions can be found for important areas of application. Examples of this are given in Chapter 10.

5. Acoustic sensors and signal processing

5.1 The importance of acoustic measurement for production automation

Acoustic testing procedures use airborne and structure-borne sound signals in order to draw conclusions on production quality or the operating state of technical products, machines and installations. Table 5.1 gives a few typical applications of acoustic measuring and recognition procedures (5.1).

Table 5.1   Applications of acoustic measurement and recognition procedures.

Application	Objective	Examples
Process automation	Control and monitoring of production processes	Drop forging Steel converters Cement mixers
Early warning of faults	Monitoring of plants and machines	Power stations Turbines
Quality control and operational testing	Assessment of the product quality	Gearboxes, Bearings, Car engines, Electric motors, Sewing machines, Tiles

With passive methods the noise made by the object itself is evaluated whereas with active methods the oscillation behaviour of a product due to artificial stimulation is evaluated. A simple example taken from the experience of everyday life is the testing of glasses by tapping. Here the resulting sound gives information on the presence of a defect.

The application of acoustic measuring methods is especially relevant to tasks where humans are employed in making the decisions. There are many such tasks in industry. Thus nearly all products of large-scale manufacture where the operation is accompanied by noise (car engines, electric motors, gearboxes, sewing machines and many others) are listened to in order to detect operational and assembly faults.

However, prolonged listening to complex noises is very tiring for humans. The demands on concentration due to constant repetition and environmental noise lead to psychological stress which lowers testing efficiency. Whereas humans are very good at making relative judgements (e.g. comparing two sounds) they have a limited capacity to make absolute judgements. They possess a bad long-term memory for non-melodic complex sounds. The application of objective measuring procedures is therefore a desirable objective in many cases for reasons of improving the quality of working life, rationalisation and increasing the efficiency of testing.

The importance of acoustic testing procedure can be seen from their considerable scope. They make it possible to detect

- production faults and damage in single components,
- assembly and opertional faults e.g. during assembly of the individual parts as well as during interaction of components and assemblies,
- excessive noise production

The question of whether an object has a fault of what fault is present is a problem of pattern recognition which will be dealt with later in more detail (Chapter 8).

A treatment will now be given of the possibilities of sound measurement as well as the principles of sound processing insofar as this relates to pattern recognition systems used in industry.

In its physical nature sound consists of mechanical oscillations of an elastic medium. A distinction is made here between airborne sound and structure-borne sound. Airborne sound is best known in daily life. Structure-borne sound is taken to mean mechanical oscillations in solid structures. A distinction will also be made between the measurement of structure-borne and airborne sound corresponding to the different measurement procedures. What is involved here is a brief summary of the essentials.

5.2   Transducers for airborne sound

Airborne sound transducers (microphones) are systems which convert sound energy into electrical energy. Usually the sound energy is converted into mechanical energy by inserting an oscillatory mechanical system (a diaphragm) and the mechnical energy then converted into electrical energy. Electroacoustic sound transducers are classified according to the type of mechanical to electrical energy conversion. The most important practically are the electromagnetic, electrodynamic, piezoelectric and electrostatic sound transducers.

Electromagnetic microphones consist of a permanent magnet with a winding and moving armature which is lined to the diaphragm (Fig. 5.1).

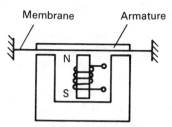

Fig. 5.1   Principle of an electromagnetic microphone.

The diaphragm is set in motion by the sound field. The alternating voltage induced in the coil winding is proportional to the speed of the membrane.

The electrodynamic microphone consists of a stationary permanent magnetic field and a conductor moving in it. This conductor is either a moving coil (dynamic microphone) or a light metal foil (ribbon microphone). The principle of the dynamic microphone is shown in Fig. 5.2.

The induced voltage in the coil in the sound field is proportional to the speed of the diaphragm movement.

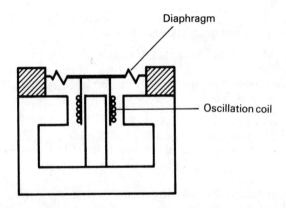

Fig. 5.2 *Principle of the electrodynamic microphone.*

With the piezoelectric principle a mechanical deformation is converted to an electrical charge. Suitable materials here are quartz, tourmaline, Rochelle salt and polycrystalline ceramics such as barium titanate or lead zirconate titanate.

The basic structure of such a microphone is shown in Fig. 5.3 using the example of a flexural resonator which is stimulated by a diaphragm.

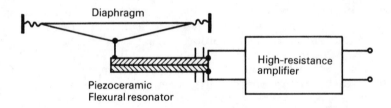

Fig. 5.3 *Principle of a piezoelectric microphone (flexural resonator).*

Electrostatic microphones are capacitors with a rigid elecrode and an oscillatory membrane electrode. The operting principle is shown in Fig. 5.4. The output voltage is proportional to the diaphragm deflection.

The capacitor microphone needs a polarisation direct current of 50 V-200 V for the purposes of linearisation. The force acting between two capacitor plates is proportional to the square of the applied voltage. The high polarisation voltage shifts the operating point to the linear region of the characteristic for small modulations.

When the microphone is exposed to sound waves its capacitance varies with rhythm of the sound and so does the current through the resistance R. The resulting drop in AC voltage is amplified by means of a high-impedance amplifier. The output voltage is proportional to the deflection of the diaphragm. The high polarisation voltage however, has practical disadvantages.

By using electrets there is no longer any need for the polarisation voltage. Electrets are materials which generate a permanent electrical field (5.2). Such a material between the electrodes of a capacitor microphone generates the required permanent field without the need for a polarisation voltage. Capacitor microphones are most often used in practice (5.3).

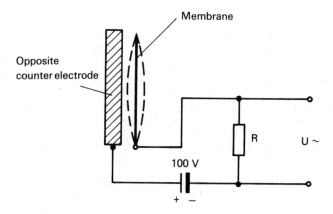

Membrane

Opposite
counter electrode

R

U ~

100 V

+ −

Fig. 5.4   Principle of an electrostatic microphone (capacitor microphone).

5.3   Characteristics and properties of airborne sound transducers

5.3.1   Directional characteristics

A distinction is made between pressure-gradient microphones and pressure micro-phones. Pressure-gradient microphones measure the difference in pressure between the front and back of the microphone diaphragm. The measuring sensitivity has a strong directional dependence as shown in Fig. 5.5a (8 character-istic).

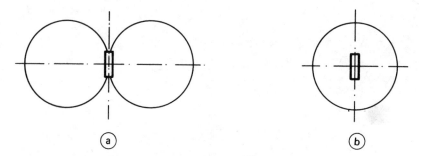

(a)   (b)

Fig. 5.5   Basic shape of directional characteristics.
(a) Pressure-gradient microphone (8 characteristic).
(b) Pressure microphone (omnidirectional characteristic).

Pressure microphones are of greater importance for sound measuring tech-nology. Here the sound waves can only act on one side of the diaphragm. In the ideal case when the sound frequency is large compared to the dimensions of the microphone the omnidirectional characteristic shown in Fig. 5.5b is obtained.

With actual microphones the sound field is disturbed by the presence of the microphone. High-frequency sound waves in particular cause a deviation from the

39

ideal omnidirectional characteristic. In order to compensate this effect the manufacturers of microphones specify so-called free-field correction curves. They describe for every 30° sound incident angle the deviation in the sound pressure acting on the diaphragm from that in an undisturbed sound field. In conjunction with the frequency response (cf. section 5.3.2) a corrected omnidirectional characteristic is then obtained, a typical example of which is shown in Fig. 5.6.

The lower the frequency the more independent is the tone pitch of the incident direction of the sound. Above the audible range (16 kHz) the directional characteristic is strongly lobar. With smaller microphones with diameters of $\frac{1}{2}$ or $\frac{1}{8}$th of an inch the directional characteristic is circular up to quite high frequencies.

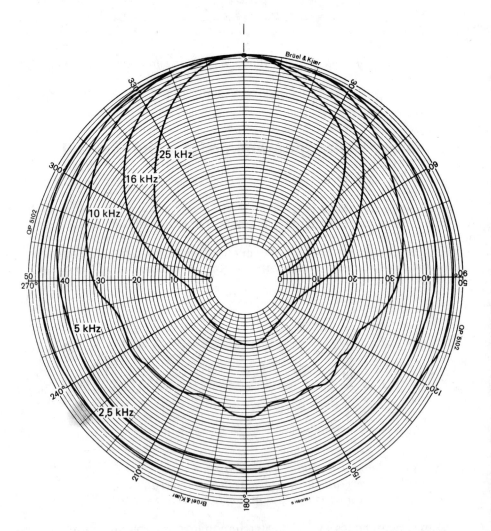

Fig. 5.6   Example of a polar diagram of the sensitivity of a 1 in. microphone (5.4).

5.3.2 Frequency response

The pressure response B of a microphone is defined as the quotient of the output voltage u to the received sound pressure p (5.4).

$$B = \frac{u}{p}$$

(5.1)

This variable gives a direct measure of the sensitivity.

Instead of the pressure response the electroacoustic index G is often used

$$G = 20 \cdot \log \frac{B}{B_o}$$

(5.2)

where B_o is the reference pressure response (10 V m²/N is a typical value) (5.5). A distinction is made between the open-circuit and the operational pressure response of electroacoustic index depending on whether the microphone voltage is measured under open-circuit or under operational closed-circuit conditions.

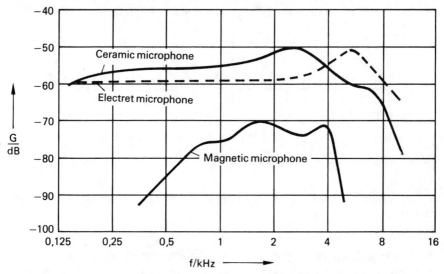

Fig. 5.7   The electroacoustic index as a function of frequency (5.4).

Typical frequency responses of the electroacoustic index are shown in Fig. 5.7 for different types of microphone. This is only a rough qualitative picture. The frequency response is very strongly dependent on the size of the microphone. The higher the frequency to be measured the smaller should be the microphone. Some typical values are shown in table 5.2 (5.6).

Table 5.2   Microphone technical data

Type of microphone	Diameter mm	Frequency range Hz	Pressure response mV/Nm⁻²	Dynamic range dB
Capacitor	24	3– 18500	50	15–145
Capacitor	12	4– 40000	12.5	30–160
Capacitor	6	4–100000	4	50–172
Capacitor	3	6–140000	1	64–176
Electrodynamic	33	30– 20000	2	10–150
Ceramic	24	3– 10000	3	30–140

41

5.3.3 Dynamic range

Apart from frequency response and directional characteristics the dynamic range is also of importance for microphones.

Microphones with a large diaphragm diameter have the best measuring sensitivity. They are especially suitable for low sound levels. Small diaphragm diameters allow high sound pressures to be measured. The operating range is therfore strongly dependent on the diaphragm diameter (Table 5.2).

Out of practical considerations the sound pressure is not expressed in microbars or Newtons/m² but as the sound pressure level p in Decibels (dB).

The sound pressure level is defined as follows:

$$P = 20 \log P_{eff} p_0) \, dB \tag{5.3}$$

Here p_0 is a reference value which is internationally fixed for airborne sound corresponding to the sound pressure at the audible threshold for humans at 1000 Hz ($p_0 = 2 \cdot 10^{-5}$ N/m²) and p_{eff} is the effective value of the actual sound pressure.

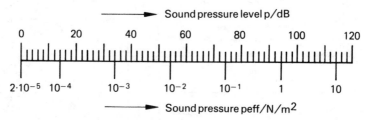

Fig. 5.8   The relationship between sound pressure and sound level.

A nomogram is shown in Fig. 5.8 giving the relationship between sound pressure and sound pressure level.

5.4   Transducers for structure-borne sound

The term structure-borne sound refers to mechanical vibrations in the solid state. The most important characteristics are the vibrational path, velocity and acceleration. These variables are related mathematically so that if one is measured the other two are known. Acceleration transducers are of greatest importance in the measurement of structure-borne sound. In contrast to displacement and velocity meters acceleration meters can be manufactured more easily and their geometrical dimensions made smaller so that the result of measurement is less distorted by their mass.

The most common acceleration transducers are piezoelectric transducers made of quartz or barium titanate. The principle most commonly used with piezoelectric acceleration transducers is shown in Fig. 5.8.

The transducer element consists of two piezoelectric plates with a weight pre-stressed by a spring attached to them. An axial acceleration causes the weight to exert a force on the piezo plates. These then act as thickness transducers.

A charge is then produced at the connections of the piezoelectric plates proportional to the acceleration. This charge is then transferred to a suitable amplifier. The amplifier can either be a voltage amplifier or a charge amplifier. Charge amplifiers are also voltage amplifiers with capacitative feedback between the output and input. The principle of a charge amplifier is shown in Fig. 5.10.

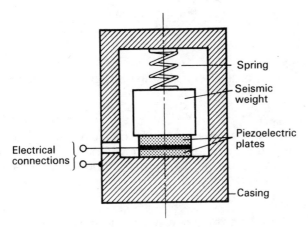

Fig. 5.9   Construction of a piezoelectric acceleration transducer (5.7).

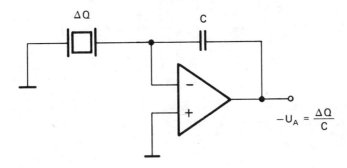

Fig. 5.10   Principle of a charge amplifier.

An advantage of the charge amplifier over the voltage amplifier is the insensitivity to changes in the capacitance of the cable.

5.5   Characteristics and properties of structure-borne sound transducers

5.5.1   Transfer factor

The measuring sensitivity or the transfer factor of an acceleration transducer is defined as the quotient of its electrical output and its mechanical input. A distinction is made between the charge transfer factor B_q

$$B_q = \frac{\text{Generated charge q}}{\text{Acceleration}} \qquad (5.4)$$

(expressed in PC/g with $g = 9.81$ m/s^2)
and the voltage transfer factor B_u

43

$$B_u = \frac{\text{Generated open-circuit voltage u}}{\text{Acceleration}} \tag{5.5}$$

(expressed in mV/g).

Typical values for the transfer factors B_u and B_q are given in table 5.3.

5.5.2 Frequency response

The operational frequency range is essentially limited by the resonance frequency of the transducer and the object of measurement as well as the rigidity of the coupling. Fig. 5.11 shows the typical frequency response of the voltage transfer factor for an acceleration transducer.

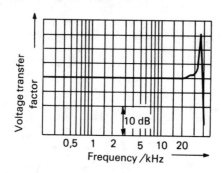

Fig. 5.11 *Frequency response of the voltage transfer factor for a piezoelectric acceleration transducer (example).*

Table 5.3 *Characteristics of piezoelectric acceleration transducers.*

Typical application	Measuring range g	B_q pC/g	B_u mV/g	Frequency range Hz	Mass grams
Standard applications	1000	10	4	2– 7000	25
Small dimensions	1000	1.5	3	10– 8000	0.5
Small accelerations	200	90	400	2– 3000	75
Shock measurements	100000	0.004	0.01	50–50000	1.5

This type of curve applies to a rigid coupling with the object of measurement. It is especially important to make the attachment of the transducer to the object of measurement as rigid as possible at high frequencies. Screw and adhesive joints are most suitable here. Coupling using a magnetic clamp limits the frequency range (5.4).

Table 5.3 lists some of the characteristics of piezoelectric acceleration transducers.

5.5.3 Directional factor

The directional factor also called the cross-sensitivity expresses the sensitivity to acceleration perpendicular to the axial direction. The cross-sensitivity mainly

44

results from irregularities in the piezo material and from incomplete contact between the piezo material and metal parts in the transducer. Normally the cross-sensitivity is 1-4% of the sensitivity in the axial direction.

5.6   Methods in acoustic signal processing

The most commonly measured acoustic variables measured in practice are the sound pressure or sound level for airborne sound and the acceleration for structure-borne sound. Fig. 5.12 shows the basic block diagram for airborne or structure-borne sound measurement with subsequent pattern recognition evaluation circuit.

The evaluation circuit usually involves the information processing and discrimination stage already indicated in Fig. 5.12.

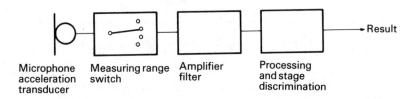

Microphone | Measuring range | Amplifier | Processing
acceleration | switch | filter | and stage
transducer | | | discrimination

Fig. 5.12   Block diagram of an acoustic recognition system.

The objective of acoustic pattern recognition systems is to use the measurement signal to assign the test object to certain categories with common properties (pattern classes).

Fig. 5.13 shows the general structure of an acoustic pattern recognition system and the design procedure.

The measurement signals generated from suitable transducers are used to extract the symbolic features which are of significance for the pattern classes.

By processing the features the classifier decides to which class k_i the test object belongs, e.g. 'product perfect' or 'product defective' or error k, error k_2, . . . error k_i.

Usually the first step is to define the error categories as shown in Fig. 5.13 on the righthand side, i.e. the objective of the system is defined.

The next step is to obtain the signal. Here the choice of the measurement location is vitally important. For instance it should not be at the node of a standing wave. Furthermore with structure-borne sound there are large transition losses at screwed joints, flanges or points of impact. It is therefore often necessary to obtain the signal in the vicinity of the place where the faults are produced. Experience shows that the final determination of the place where the faults are produced can only be made as the result of signal analysis. The selection of suitable measuring points is an economic operation and makes testing more efficient than carrying out the complicated processing of signals which are obtained at less favourable locations.

Reference should be made here to a special problem which occurs when auto-mating acoustic pattern recognition. The individual sound categories have to be first formed by specially trained test personnel. The set of objects for this should consist of as many representatives as possible. In addition several test personnel should be involved so that the final assessment depends on the result of several test personnel. The result of this subjective assessment then forms the basis for the objective measuring procedure to be developed.

If a system has been developed on the basis of these features then the objects to be

45

tested are classified in a traditional stage objectively by the system as well as subjectively by the test personnel. A comparison of the results makes it possible to improve the efficiency of the automatic recognition system.

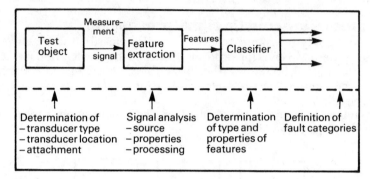

Fig. 5.13   Pattern recognition system for acoustic tasks and its design (5.1).

As a rule the main cost and effort in developing an acoustic recognition system is not the hardware but the software. The software covers the subjective assessment, the signal analysis and the selection of the classifier and statistical protection.

While classification does not present any basic difficulty feature extraction is highly problematical. There is no closed procedure here which is generally applicable.

It has been found that adaptive signal processing provides a versatile method of feature extraction (5.1, 5.8). Here the signal properties are analysed taking into consideration the physics of the signal origin. Knowing the signal properties, processing methods are applied which are especially adapted to these properties.

Chapter 8.2.1 gives more details on the method of adaptive signal processing. However some general information will be given at this juncture (5.8).

Deterministic movements (i.e. capable of being described analytically) of machine parts predominantly give rise to deterministic noises. For instance uniformly rotating gears or turbines (5.9) as well as the rolling movement of roller bearings (5.10) generate almost periodic vibrations. They are described as a tone when one frequency is dominant (5.11). Usually rotation movements as well as periodic translation movements and shock pulse trains are associated with harmonics n.f.$_1$ of the fundamental frequency f_1 so that full-sounding sounds are produced. These deterministic vibrations are described by an explicit mathematical relationship. In this way the individual frequencies can be determined from the geometry and the rotational speed.

Mechanical impacts and blows produce single vibration processes. They originate from different forms of energy release. The variation over time of the amplitudes of such a transient vibration is usually determined by the eigen frequency and the damping constant (cf. section 8.3.1.3). The impact pulse train can be periodic or random. Eccentric presses, defective brushes on electric motors or the internal combustion engine generate periodic sequences. Hammer blows of a forging press or tapping ceramic workpieces in the sound test are on the other hand irregular in nature.

Streaming liquids in pipes, turbulences, cavitations in fittings and valves, friction in friction bearings and sliding clutches produce stochastic, usually broad-band noise. Expanding gas produces a noise with the same output for all frequencies in the audible range (5.12). Narrow-band noise is produced for instance when gases emerge through jets from resonance chambers. The whistling in the blowing process during steel production is a result of this (5.12). In general these stochastic processes retain their character over a long time and are then assumed to be stationary.

Non-stationary processes change their statistical properties during the observation period (cf. section 8.3.1.1). Non-stationary processes are mainly the result of instabilities and other changes in the physical mechanism or arise when the same input process gives rise to different output processes. An example is the squeaking during the shaping of metals which is the result of cutting.

This list shows that the origins of sounds and the sounds themselves can be varied. In addition it should be noted that the noise which emanates from a machine at a certain location does not only depend on the machine but also on its surroundings. Resonances often amplify an originally insignificant vibration which then becomes a main source of sound.

6. Tactile transducers and signal processing

6.1 Objective

Investigations in the assembly sector have shown that joining operations with narrow tolerances have to be carried out at over 50% of the workplaces (6.1) Fig. 6.1 shows the six joining operations which are carried out most frequently.

Even when the positions of the parts to be joined are fixed exactly there can still be difficulties with manipulation. For instance difficulties can arise when inserting a bolt into a hole with little play or turning a screw in a thread due to twisting or a defective thread despite the position being known and in some cases the operation cannot be carried out.

If such processes are to be automated then tactile sensors are necessary which allow sensitive joining or parts.

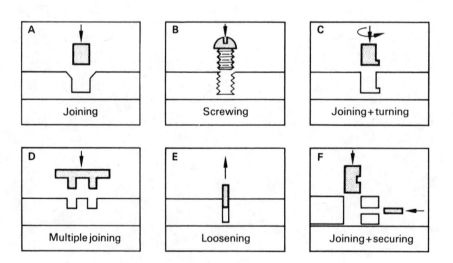

Fig. 6.1   Joining operations in assembly.

Another important task in the tactile area is the tracking of edges on objects. This curve tracking occurs for example in industry when deburring work-pieces.

With joining as well as with curve tracking, forces and torques are generated which can be used to determine and control the movement.

The transducer of such a tactile sensor will thus be as a rule a force or path transducer.

The block diagram of a tactile sensor for joining operations or path tracking is shown in Fig. 6.2. The signal from the transducer is processed in a subsequent stage according to a movement strategy and used to correct the movement process.

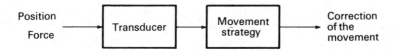

Fig. 6.2   Principle of a tactile sensor for joining operations or path tracking.

The functions of tactile sensors are however not just limited to joining parts or path tracking. Corresponding to humans being able to recognise objects by the sense of touch alone tactile sensors can also be used to solve recognition tasks. In this case an object is recognised on the basis of appropriate signals which are derived from the object geometry during positive locking. This can be done for example as shown in Fig. 6.3 with the position determined by moving pins integrated into a gripper and adapting themselves to the object.

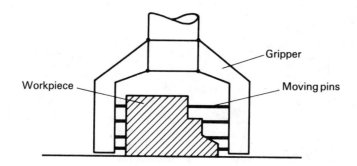

Fig. 6.3   Diagram of a tactile sensor for recognition tasks (gripper with adjustable jaws).

With tactile recognition tasks the variable to be determined is usually a change in position. Because of the close relationship between path and force measurement (Section 6.3) the transducer as a rule also consists of a path or force measuring system.

The block diagram for a tactile sensor for recognition tasks is shown in Fig. 6.4. It corresponds to that of optical and acoustic sensors.

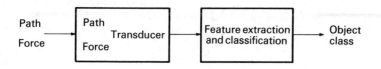

Fig. 6.4.   Principle of a tactile sensor for recognition tasks.

6.2 Analysis of the joining operation

Since most tactile tasks in assembly are joining operations we shall examine these more closely.

Fig. 6.5 shows the process of inserting a bolt. First one side of the hole is touched (one-point contact). By tilting the bolt strongly to the axis of the hole and inserting the bolt futher into the hole a two-point contact is made with the risk of jamming (6.2).

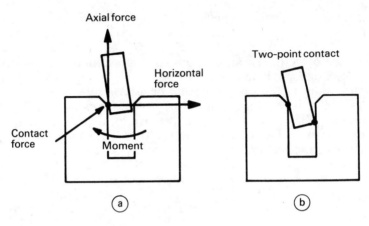

Fig. 6.5   Joining operation.

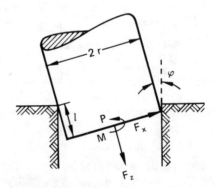

Fig. 6.6   Derivation of the joining conditions.

Fig. 6.6. will be used to derive the joining conditions more exactly. F_x, F_z and M are the forces and moment acting at the point P of the bolt; 1 is the penetration depth, Q is the joining angle of the bolt and μ the coefficient of friction (e.g. for steel on steel: $\mu = 0.1$). Assuming a low joining speed and a rigid bolt the following condition can be obtained for the joining operation (6.3):

$$\frac{M}{F_z \cdot r} + \frac{F_x}{F_z} \leqslant 1.$$

$$\lambda \sin(\varphi) \qquad \frac{\lambda \sin(\varphi)}{(\lambda + 1)\mu} \qquad (6.1)$$

This is valid for the region $\lambda > 1$ with

$$\lambda = \frac{1}{2r\mu} \qquad (6.2)$$

Fig. 6.7 shows the two regions where joining is possible and where the part can jam as a function of the determining parameters. According to this for the joining operation to be feasible the reaction forces and moments not acting in the joining direction must be small.

If jamming occurs during a joining operation then a corrective movement must be made to release the two-point contact so that joining is possible again. This principle is often used in the movement strategy of a tactile sensor engaged in joining operations. The measurement of the resulting forces and moments is used to control the corrective movements which are made to reduce the interfering reaction forces and moments.

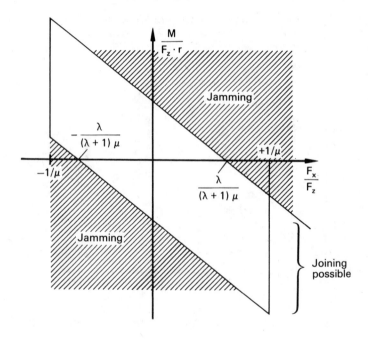

Fig. 6.7   Joining and jamming regions (6.3).

This leads to the basic principle for the joining device with tactile sensor shown in Fig. 6.8.

The reaction forces during a joining operation are not transferred in a rigid manner but by an elastic arrangement of the gripper on the joining device. The force transducers for measuring the reaction forces are also coupled to this elastic structure. A typical curve which in principle looks the same for horizontal and axial forces is shown in Fig. 6.9. A steep rise in the force after the two-point contact has been made contains the risk of jamming. The corrections to the joining movement should be made so that the rise in the reaction force is limited or reduced.

51

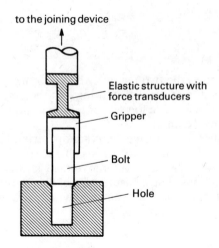

to the joining device

Elastic structure with
force transducers

Gripper

Bolt

Hole

Fig. 6.8   Principle of a joining device with tactile sensor.

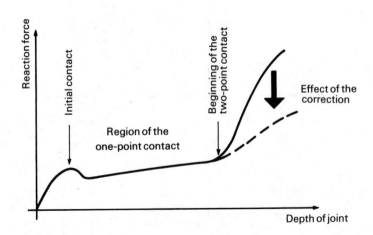

Reaction force

Initial contact

Beginning of the
two-point contact

Effect of the
correction

Region of the
one-point contact

Depth of joint

Fig. 6.9   Basic curve for the reaction forces during a joining operation (6.2).

6.3   Force and path transducers

For the sake of completeness a brief account will now be given of some of the
methods for measuring force and path. This material is adequately handled in the
general measuring technology literature, e.g. (6.4, 6.5).

If a force acts on an elastic body then it will be deformed. The magnitude of the
force can be determined by measuring the deformation. Here so-called elastic
structures are used the deformation of which is defined as a function of the acting
force.

The best known elastic structures are bending beams, triangular, ring, diaphragm
and tube springs. The following procedures are usually used for force and path
measurement in conjunction with such elastic structures.

The most common are the wire strain gauges where a mechanical deformation causes a change in resistance (piezoresistive effect).

Transducers are also often used where the force is measured inductively. The deformation of the elastic structure moves an armature which alters the inductivity of a coil.

Potentiometer gauges are especially suitable for large measuring paths. Here the potentiometer tap is altered as a function of the deformation.

Recently elastomers have opened up new areas of application for tactile sensors. They possess the ability to vary their resistance as a function of the pressure (piezoresistive effect) (6.6, 6.7).

Fig. 6.10 shows a typical curve for the electrical resistance of elastomers as a function of pressure. The value of the resistance varies from 10MOhm for the unloaded state to approximately 10 Ohm for a load of 15 N/cm².

The thickness of the elastomer material is 0.1 – 0.5 mm and is therefore good for covering metal surfaces. Less common are magneto-elastic structures in which the pressure dependence of the permeability on the force or position is exploited.

Force transducers are also widespread which use the piezoelectric effect. In this case a charge is generated in certain materials through the action of the force. The best known piezoelectric material is quartz. Because of the small construction height these gauges are especially suited for multicomponent measurements.

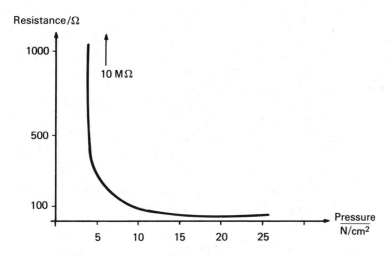

Fig. 6.10   Electrical resistance of elastomers as a function of pressure.

Certain polymers are suitable for exploiting the piezoelectric effect for transducers in tactile sensors – similar to the use of piezoresistive plastics (6.8). With some polymers, e.g. polyvinylidene fluoride a charge is generated by pressure. An advantage of these materials is their high elasticity which means for instance that they can make a good fit with the geometrical shape of an object being gripped for recognition. The generated charge is then a function of the object geometry.

Torques are nearly exclusively measured by determining the twisting of an elastic structure usually tubular in shape. In principle all the above-mentioned procedures are suitable for measuring this twisting. However wire strain gauges and inductive transducers occupy a certain privileged position.

7. Microprocessors in sensor applications

Up until now in the diagram shown in Fig. 1.2 of a sensor system we have been discussing the transducers which supply the measurement signal. We shall now briefly consider the implementation of the information processing and discrimination stage. For this stage a computer system suggests itself which can make the necessary decisions.

Electronic computer systems were previously employed for complex tasks: for processing large volumes of data in commerce, for controlling large-scale processes or for solving complex problems in science and industry. This is the reason why computers were very expensive and occupied a lot of space. This situation began to change with the advances being made in semiconductor technology. This made it possible to make minicomputers which although they are less productive and convenient than large installations are much superior in terms of cost and space.

The most recent step in this direction had led to the microcomputer with several thousand components on a semiconductor chip. The microcomputer is therefore the slimmed-down version of the minicomputer. It also has a smaller number of commands. Fig. 7.1 shows a diagram of the evolution in size of computer systems.

The central unit of a microcomputer is called the microprocessor. A micro-computer can be assembled from a microprocessor by supplying it with additional memories and an input/output unit. Because of this ease of conversion and the overlapping of the architecture and functions the boundaries between microprocessor and microcomputer have become very hazy.

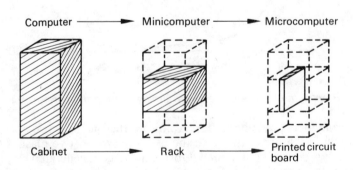

Fig. 7.1   Evolution and size of the central unit in computer systems.

With this microprocessor first introduced in 1972 all the functional units can be accommodated in a few integrated modules. Correspondingly it is now possible to solve sensor problems with the methods and advantages of computer technology which would have been impossible up until now for reasons of cost.

In principle it is possible to solve the same problem with a hardwired logic circuit as with a programmable computer.

Due to the sequential execution of the program instructions in a microprocessor it takes longer to process the information in a microprocessor than in hardwired logic circuit. However the great importance of microprocessors lies in the variability of program and function sequences. As a rule systems which rely on hardwired logic are more advantageous when very high speeds are required. If however a large degree of flexibility is desired then processors are more suitable because of the programmability, i.e. the ease in making subsequent alterations when changing over to new tasks.

A microprocessor is therefore highly suited for use in a sensor. It combines a discrimination and processing capacity with the condition of flexibility important for sensors.

The following gives a short account of the basic organisation of a microprocessor system. Further literature can be found in (7.2 – 7.9). A general introduction to computer structures can be found in (7.10).

All microprocessors are built from similar functional units although their internal structure differs from manufacturer to manufacturer. The principle of a microprocessor is shown in Fig. 7.2.

The basic units of a microprocessor system are the

- microprocessor unit,
- read-only memory for the program commands
- random access memory for the results and variable data
- input/output unit

The individual system components are connected by a common data channel called the bus (7.11). This bus has all the necessary lines for the data, address and control signals. Apart from the control signals it is designed in such a way that all

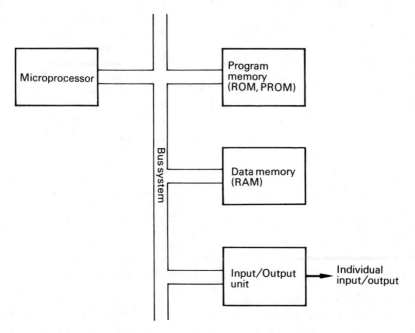

Fig. 7.2 Principle of a microprocessor system.

signals on the lines can move in both directions and be available at all functional units at the same time.

The following gives a brief account of the function and operation of the system components shown in Fig. 7.2. A knowledge of their functioning is necessary for understanding the system implementations considered in Section 10.

The microprocessor represents the central unit (MPU: microprocessor unit, CPU: central processing unit) which contains the control unit and the arithmetic unit. Fig. 7.3 shows a possible basic organisation for this central unit.

The program counter is a buffer for storing the address of the next command which is executed in the program sequence. Some commands alter the content of the program counter so that the following command is not the next in the program sequence. This makes it possible to jump from one point in the program to another depending on the conditions.

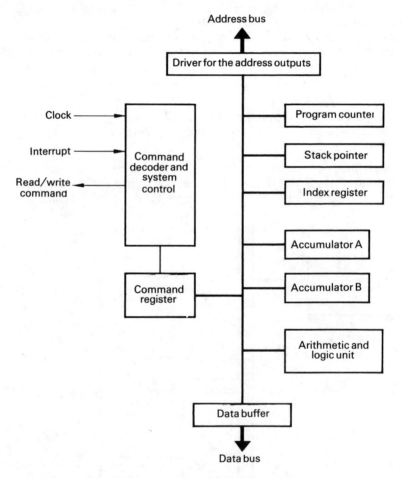

Fig. 7.3   Organisation of the central unit of a microprocessor system.

A part of the random access memory is reserved for the stack. It mainly serves to handle interrupts and subprograms more easily. The stack pointer keeps the address of the next free place in the stack.

The index register is mainly used for changing addresses. A command contains the operation code (load, add, subtract, etc.) and an address at which the data to be operated on is located. Thus a command can take up three bytes: one byte for the command itself and two bytes for the address. This address is modified by using the index register.

The accumulators contain one of the operands and with many commands also the result.

The arithmetic and logical unit (ALU) is the actual calculator where the arithmetic functions such as addition, subtraction, comparison, bit scan and logical functions such as AND, OR, NOT are carried out.

The command register stores the operation code which is retrieved from memory. After decoding internal signals are generated from this for controlling the arithmetic and logical unit.

The next component of a microprocessor system (Fig. 7.2) is the read-only memory for the program commands. Read-only memories (ROMs) have fixed binary information which cannot be altered. It is also not lost in the event of power failure. Apart from mask-programmed ROMs which are programmed by the manufacturer once and for all there are also memories which can be programmed by the user (PROMs).

In addition the EPROM (erasable PROM) can be erased again with the help of UV light. It can be reprogrammed with a special programming device and used for other tasks.

The next component of a microprocessor system is the data memory (Fig. 7.2).

Random access memories (RAMs) are used for storing variable information in which every memory location can be addressed. These memories can be read, written to and erased.

With RAMs a distinction is made between dynamic and static modules.

The static RAMs consist of integrated flip-flops. These days microcomputers are mainly provided with such memory elements.

Dynamic RAMs use capacitors instead of highly integrated flip-flops for information storage. Becasue of the leakage currents from these capacitors the loss of charge must be compensated in intervals of a few milliseconds (refresh).

Additional circuits are necessary for this.

RAM memories belong to the volatile memories where the information is lost in the event of power failure. With sensor systems where the information must be retained buffer batteries are used to circumvent power failures.

The microprocessors which are on offer today can be classified as follows: 4-bit microprocessors (e.g. Intel 4040, Texas TMS 100), 8 bit microprocessors (e.g. Intel 8080, Motorola 6800, Fairchild F8), 16 bit microprocessors (e.g. PACE, TMS 9900, micro NOVA). The 8-bit microprocessors are most commonly used – also at the present time for sensor systems.

However for picture processing 16-bit microprocessors are looking increasingly promising. They are much better suited for dealing with large quantities of data.

A development system makes it easy to develop individual user programs. A corresponding hardware and software makes it possible to develop programs in a similar manner to mainframe computers. However there is not the same programming facility. The program development is carried out in a processor-specific assembler language. Although there are addition and subtraction instructions in the instruction set direct multiplication and division are not yet possible and have to be carried out by subprograms.

Overall a microprocessor system presents the ideal flexible system for realising the necessary intelligence for sensors by virtue of the low costs and the favourable dimensions. The revolutionary developments in the area of microprocessor technology will also considerably expand the possibilities of sensor systems in the future.

8. Industrial pattern recognition methods

Humans are well equipped to recognise and discriminate between different objects and sounds. This recognition and discrimination ability is characterised by a certain intelligence.

With high levels of automation in industry the recognition tasks carried out by humans have to be carried out by corresponding sensor systems.

The following sections deal with pattern recognition methods giving special attention to their technical viability and their application in sensor systems.

8.1 Basic concept of a pattern recognition system

The task of pattern recognition is the assignment of patterns (workpieces, sounds) to certain categories, e.g. workpiece 1, workpiece 2 or noise too loud, normal noise.

The procedures which are used for the solution of these tasks are adapted from procedure in statistics, statistical communications theory, linguistics, theory of automata, information theory as well as control and systems theory. Many working systems in automatic pattern recognition are characterised by high development costs. However for extensive industrial application only those pattern recognition systems are suitable where the technical implementation bears a reasonable relationship to the economic benefits achieved. It follows from this that there is no universal pattern recognition system – only an individual solution to a certain problem area or field of application. This means that a system is developed for instance only for sound recognition or only for recognising workpieces. Both the theoretical and practical prerequisites are missing in order to develop a universal system which like the human system is able to handle all possible types of patterns.

Since the mathematical methods available often lead to no practical solution heuristic methods assume special importance in pattern recognition. These methods are those which appear intuitively reasonable and promising but cannot be derived mathematically. Heuristic methods are justified when the solution has been proved technically feasible.

Fig. 8.1 shows the principle of a pattern recognition system.

The patterns are converted by a transducer (television camera, microphone) to a suitable form of processing. The transducer already carries out basic preprocessing and information reduction. Thus for instance the colour picture of the environment is converted by a black and white television camera to a grey-tone picture.

The signals generated by the transducer are brought into a suitable form for further processing by a preprocessing stage. This preprocessing involves the suppression of interference signals as well as the extraction of features which give the best possible description of the pattern. This preprocessing and feature extraction stage transforms the transducer signals using suitable selected features

59

into a feature vector c whose components describe the pattern. In the general case the feature vector can be represented as a n-dimensional column vector (8.1, 8.2, 8.3).

$$
C = \begin{bmatrix} C_1 \\ C_2 \\ \cdot \\ \cdot \\ \cdot \\ \cdot \\ C_n \end{bmatrix}
\tag{8.1}
$$

This feature vector forms the basis for the discrimination to be made in the subsequent discrimination stage. The characteristics of the individual categories are stored in this discrimination stage. The result of the discrimination is a category name, e.g. during a recognition process: workpiece 1, workpiece 2, . . ., workpiece n. Thus a pattern recognition system contains a feature extraction stage in which the transducer signal is processed in such a way that the classification can be made in the subsequent discrimination stage.

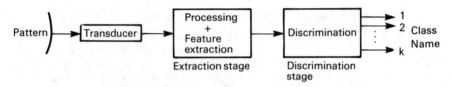

Fig. 8.1   Principle of a pattern recognition system.

The most difficult problem here is to find suitable pattern features. There is no closed procedure which gives the best features for a certain class of problems.

On the other hand the problem of classification is clearly defined: a given feature vector has to generate a code for the appropriate class as an output variable. This problem has been solved theoretically.

The most important procedures for preprocessing and classification in industrial application will be described in the following two Chapters 8.2 and 8.3.

8.2   Preprocessing and feature extraction

The result of preprocessing and feature extraction is a feature vector which enables the simplest and most reliable decision possible to be made on the membership of a given class. Here it is advantageous to adapt the procedure to the special conditions of the task in question. In this case one speaks of adaptive signal processing.

In acoustics the signals being investigated are usually functions of time or frequency. On the other hand in image processing spatial variations in intensity (spatial frequency) are involved. The following gives a selection of procedures as they are used by preference in one of the named areas. However it should be emphasised at this point that the individual procedures are by no means restricted to the specified applications. For instance average values need not only be calculated from time functions.

However for practical applications in sensor systems it has proved advantageous to treat the same processing methods in acoustics separately from those in picture processing. This results from the marked differences in the objectives, signal properties, auxiliary conditions and significance of the individual variables. This can be especially clearly seen with the two examples presented of homomorphous filtering (Section 8.2.1.8) or with the correlation procedure (section 8.2.1.3), which crops up again with image processing as a possible method of template comparison (Section 8.2.2.2).

8.2.1 Acoustic signal processing

With adaptive signal processing in acoustics a prior knowledge of the signal origin is exploited to gain information on quite specific signal properties (8.4). Typical examples of such signal properties are the harmonic content, cycle dependence, the extent of narrow or broad bandedness, the degree of modulation, etc. Knowing the signal properties, procedures are applied which specifically relate to these properties. An unadaptive procedure is for instance the application of a spectral analysis irrespective of the type of signal. An amplitude modulation can then not be recognised in the power density spectrum. Fig. 8.2 shows some connections between signal origin, signal properties and the most suitable methods for acoustic processing (8.4). The processing methods employed originate from systems and

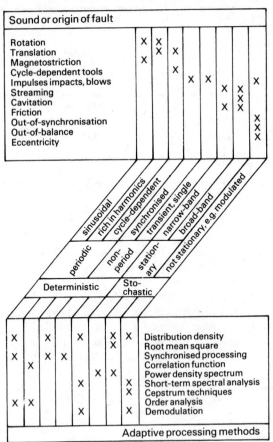

Fig. 8.2 Signal origins, signal properties and adaptive acoustic processing methods (8.4).

communications theory. These connections are based on a number of industrial examples; other connections are however conceivable and possible.

The following sections give some detail on the processing methods specified in Fig. 8.2.

As a rule acoustic signals are described in terms of a single parameter (time, frequency). With tasks in acoustic pattern recognition only in rare cases are deterministic signals involved which can be defined analytically in a unique manner. Feature extraction usually involves stochastic, i.e. random signals which have the characteristic that the observer cannot recognise any regularities in their behaviour. Only probability statements can be made on such processes and these never apply to the single cast but to an entire assembly of identical processes. This means that a very large number of signals have to be evaluated within an ensemble of identical systems. Fig. 8.3 shows three sample functions also called representatives of the parameter t, time.

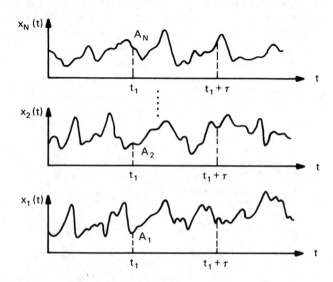

Fig. 8.3   Sample functions of a stochastic process.

The total number of sample functions forms a stochastic process. If for example in practice one wishes to determine the statistical properties of test objects based on the airborne sound generated then there must be a large number of identical test objects available so that enough sample functions are represented. The characteristic variables of the stochastic process (8.5) can then be determined from the ensemble of sample functions which is very long-winded.

With a stationary stochastic process all characteristic variables are independent of time. Thus for instance the average value to be calculated over the ensemble x_e (Fig. 8.3).

$$\bar{x}_e(t_1) = \lim_{N \to \infty} \frac{1}{N} \sum_{K=1}^{N} A_k(t_1)$$

$$(8.2)$$

is time dependent for a stationary process, i.e.

$$\bar{x}_e(t_1) = \bar{x}_e(t_1 + \tau)$$

$$(8.3)$$

The statistical variable have also to be determined from a large number of sample functions assuming that the process is stationary. It would be much more advantageous to calculate these variables from the variation of a single sample function over time. This would correspond to actual practice where as a rule only a few samples are available.

For a special class of stochastic processes the average value over the ensemble (equation (8.2)) is actually identical with the time average of a single sample function. These are the ergodic processes.

It is often difficult or impossible to test whether a process is ergodic. In order to make the best use of the advantages of ergodic processes in practice the so-called ergodic hypothesis is made. One assumes that the process is ergodic and checks on the basis of the results whether this assumption is justified or not.

The physical justification for ergodicity is that a stationary stochastic process in a single occurrence runs through all amplitudes with the same probability as would be obtained at a given point in time with a large number of identical systems.

The following considerations assume stationary and ergodic processes with characteristic variables taken from the time averages.

However it is just in acoustic quality control where non-stationary processes often occur. For example the objects can heat up during the testing operations so that the behaviour is no longer stationary.

8.2.1.1 Average value and dispersion

If $x(t)$ is the time function of a process then the time average value which is identical to the ensemble average for an ergodic, stationary, random process is given by

$$\bar{x}(t) = \lim_{T \to \infty} \frac{1}{T} \int_0^T x(t)\, dt \tag{8.4}$$

The infinitely long averaging time required theoretically is obtained in practice by a sufficiently long finite averaging time. One then speaks of the short-term average value (8.6).

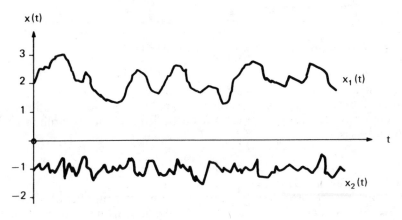

Fig. 8.4   Average value and dispersion signals.

Fig. 8.4 shows the time variation of two signals. It can be seen that the signal $x_1(t)$ makes relatively large fluctuations from the average value 2 whereas the signal $x_2(t)$ makes smaller fluctuations from the average value -1. These fluctuations are characterised by the variance σ^2:

$$\sigma^2 = \lim_{T \to \infty} \frac{1}{T} \int_0^T \{x(t) - \bar{x}(t)\}^2 \, dt \tag{8.5}$$

After separating off the constant part ($x(t) = 0$) the variance is equal to the square of the effective value x_{eff} of the varying components of the signal (8.6) and is then defined by the relationship.

$$\sigma^2 = x^2_{eff} = \overline{x^2(t)} = \lim_{T \to \infty} \frac{1}{T} \int_0^T x^2(t) \, dt \tag{8.6}$$

where $x^2(t)$ is the root mean square proportional to the power. The theoretically infinitely long averaging time must also be replaced in practice by a finite averaging time.

A measurement circuit for the variance is shown in Fig. 8.5. The square-law circuit connected after rectifier approximates with the help of the diode network (D_1, D_2 and D_3) in polygon fashion a quadratic current-voltage characteristic (8.8-8.11).

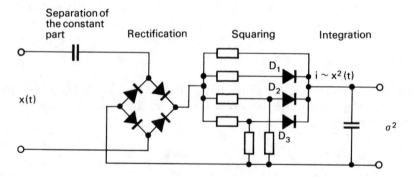

Fig. 8.5   *Measurement circuit for the variance (8.7).*

The calculation of the root mean square value from the sound signal according to equation (8.6) gives the power level of the signal in acoustic applications. In many cases the power level is a significant acoustic feature, for example in friction or grinding noises.

8.2.1.2   Synchronised averaging

Many machine processes have the characterstic that a number of events occur in the rhythm of an operational sequence. For instance with an internal combustion engine this rhythm consists of the four cycles of suction, compression, ignition, expansion. There are similar causal relationships for the sounds of forging presses, embossing and stamping machines or with sewing machines and looms. The process is also here rhythmically structured: no-load running, engaging, operating cycles make it possible to expect that the noise and the vibration of the machine will be influenced by the working rhythm.

In such cases an analysis of the noise or the structure-borne sound can follow the operating sequence. Time marks can be derived which enable the evaluation to be matched to the machine cycle. The signal processing can be synchronised by the time marks to include several cycles. Examples of this can be found in the early warning systems for Diesel engines (8.12), turbomachinery (8.13) and precision devices (8.14, 8.15) as well as in the localisation of local damage on gears (8.16) (Section 10.2.1).

The procedure can also be applied to transient processes such as during a forking stroke or when tapping ceramic parts. The sound signal of such processes can be described by one or more decaying sinusoidal vibrations as follows (8.4):

$$x(t) = \begin{cases} A \cdot e^{-Dt} \cdot \sin \omega t & \text{for } t \geq t_o \\ \\ 0 & \text{for } t < t_o. \end{cases} \qquad (8.7)$$

The damping constant D and the eigen frequency are the characteristic features of the process.

If the period of observation is limited to a single pulse then the sound is not stationary. If however the observation covers a sequence of several pulses, e.g. with gear noises the sequence of impacts of tooth profiles on meshing then the sound is stationary. Such stationary sounds contain a periodic signal component with a stochastic component superimposed on it. A synchronous averaging which is synchronised to the pulse of the sound and performed over the duration of the machine cycle eliminates the interfering noise components. Since most machine operations are coupled with rotational movements the averaging can be synchronised with the rotational movement. In this connection a mark is made on the rotating part and the averaging initiated by scanning it. A practical example will be considered in section 10.2.1.

The synchronised averaging of a time function is especially efficient when performed with the help of a digital computer. Fig. 8.6 gives an example of a signal curve which based on a synchronising pulse S is recorded anew with each revolution of the machine. Within each cycle the signal is scanned at equidistant

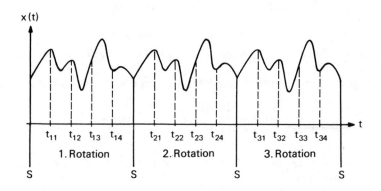

points in time. The averaging is carried out by adding the amplitude at the same scanning times in different cycles and dividing by the number of cycles considered. Thus for instance for each of the first scanned values with 3 cycles.

$$\bar{x}_1 = \frac{1}{3}\left[x(t_{11}) + x(t_{21}) + x(t_{31}) \right] \tag{8.8}$$

If during a cycle N scanning values are obtained and the cycle is run through M times then the synchronous short-term averaging generated the values (8.13).

$$\bar{x}_i = \frac{1}{M} \sum_{i=1}^{M} x_{ij} \quad j = 1, \ldots N \tag{8.9}$$

The rotational synchronous averaging reduces the noise components of the signal by a factor of $\sqrt{M}$ (8.13). A further advantage of rotational synchronous scanning is that the feature vector $\bar{x}_i$ is independent of the cycle time and so unaffected by any fluctuations.

8.2.1.3 Correlation procedures

The linear and quadratic averages considered up till now are only numbers and not functions from which one could possibly expect information on the structural properties of the signal. In this connection it is worth forming the time average of all products of function values which are separated by τ time units. This means that the average values is a function of the time difference τ which is described by the correlation function. It provides information on the extent of a structural relationship between a signal and the same signal separated by a time τ.

If one now analyses a signal $x_1(t)$ by determining its structural similarity to itself at different time displacements then the auto-correlation function is obtained:

$$k_{11}(\tau) = \lim_{T \to \infty} \frac{1}{2T} \int_{-T}^{+T} x_1(t) \cdot x_1(t+\tau)\, dt \tag{8.10}$$

In a corresponding manner the cross-correlation function is obtained for two different signals.

$$k_{12}(\tau) = \lim_{T \to \infty} \frac{1}{2T} \int_{-T}^{+T} x_2(t) \cdot x_1(t+\tau)\, dt \tag{8.11}$$

It expresses to what extent the signal $x_1(t)$ is similar to the delayed signal $x_2(t+\tau)$.

The integral is often approximated by a sum for digital calculations (8.13):

$$k(\tau) = \frac{1}{N} \sum_{n=1}^{N} x(n \cdot \Delta t) \cdot x(n\Delta t + \tau) \tag{8.12}$$

Here N is the total number of values of the time function x(t) successively measured at equal time intervals.

The correlation technique makes it possible to separate different frequency components and in particluar to extract periodic signal components. If a signal is to be filtered from a mixture of signals then the cross-correlations of this signal which have say been stored gives a maximum for the signal in the mixture which is proportional to the required signal for a certain time displacement.

Formally correlation analysis consists of the operations time displacement, multiplication and averaging of two time functions. The processing stages are shown in Fig. 8.7 in diagram form. The position of the selection switch determines whether auto or cross-correlation functions are obtained.

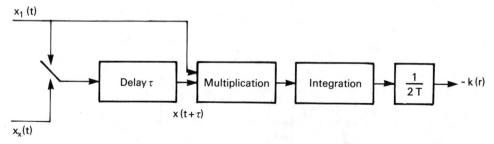

Fig. 8.7   Block diagram for determining the correlation function.

Fig. 8.8 gives some examples for the auto-correlation functions of periodic and nonperiodic signals. The correlation function of the periodic signal shows that although the shape of the signal changes all important properties such as amplitude, frequency and initial phase remain the same.

8.2.1.4   Tracking filter method
A special correlation procedure is the tracking filter technique which can also be applied with advantage to machines with rotating parts (8.4). The method scans the entire range of the excitation frequency. The sound signals consist of a tone component with fundamental frequency f_1 proportional to the rotational speed and a noise component. Examples are the wailing of turbines, the ringing of electric motors or the noise of gears.

For example the quiet running of a machine can be evaluated using the power of the frequency f_1 . For adaptive signal processing – the narrow-band selection of the frequency f_1 a harmonic reference signal is derived with frequency coupled to the rotational or excitation frequency (Fig. 8.9). The reference signal r (t) and a signal p (t) phase-shifted by 90° are multiplied with the signal x (t) being investigated. Only the constant terms resulting from the multiplication are of interest. The other sum and difference frequency components are suppressed in the low-pass filters with limit frequency f_g. Of the superimposed noise only the narrow-band component of the band width $2f_g$ is effective which is centered around the frequency f_1. The quadratic sum of the two components is proportional to the power of the harmonic.

An application example of this technique is given in section 10.2.1.

8.2.1.5   Demodulation
Many acoustic signals are amplitude-modulated. For instance typical noise impressions such as 'hoarse', 'crackling' or 'whining' indicate amplitude modulation. Fig. 8.10 shows the signal of a gearbox sound as an example of an amplitude-modulated process (8.4).

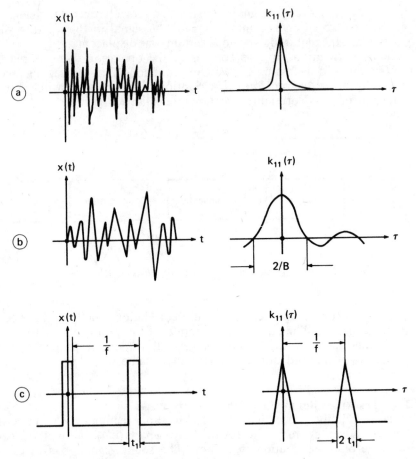

Fig. 8.8   *Examples of time functions x (t) and associated auto-correlation functions $k_{11}(\tau)$ (8.13).*

(a) Broad-band noise
(b) Band-restricted noise (band width B)
(c) Periodic signal (pulse train)

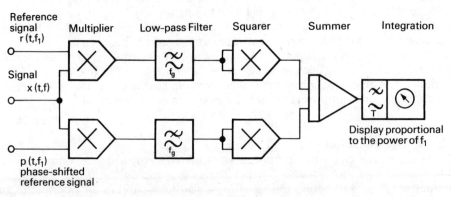

Fig. 8.9   *Principle of the tracking filter method (8.4).*

In such a process a signal is connected to the modulating function by multiplication. The objective here is to characterise the carrier modulation process.

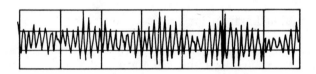

Fig. 8.10   Time variation of an amplitude-modulated signal.

An adaptive method for processing amplitude-modulated signals is demodulation. The demodulator consists of a rectifier followed by an averaging filter. The rectifier is a non-linear operation. The averaging filter is a linear network realised by a low-pass filter.

Fig 8.11 shows an arrangement for extracting characteristics when processing amplitude-modulated stochastic signals. The signal reaches the limit frequency f_H on a high-pass filter. This suppresses low-frequency interferences e.g. machine vibrations. A full-wave rectifier is used for the rectification. The subsequent demodulator low-pass filter with limit frequency f_L suppresses the carrier ripple. The result is the envelope which joins the signal peak values. The following characteristics can be obtained from the envelope (8.4):

– periodicities over time,
– a modulation factor in the amplitude range (it describes the variation in amplitude of the envelope),
– a spectral line of the modulation frequency in the spectral power density (cf. Section 8.2.1.7).

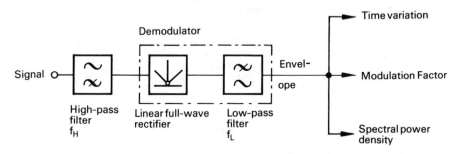

Fig. 8.11   Feature extractor for amplitude-modulated signals (8.4).

8.2.1.6   Amplitude spectrum

Formally an analytically defined time function $x(t)$ can be converted to the frequency range by the Fourier transform $F\{x(t)\}$. One then obtains the spectral function $X(\omega)$.

$$X(\omega) = F\{x(t)\} = \int_{-\infty}^{+\infty} x(t) . e^{-j\omega t} dt \qquad (8.13)$$

The associated inverse transformation is:

$$x(t) = F^{-1}\{X(\omega)\} = \frac{1}{2\pi} \int_{-\infty}^{+\infty} X(\omega) e^{+j\omega t} d\omega \qquad (8.14)$$

where $\omega = 2\pi f$ is the angular frequency.

Fig. 8.12 shows two time functions x (t) and the associated amplitude spectra as examples. These were taken from the airborne sound picked up by a microphone from intact and damaged model turbines (8.13). The damage involved bending of one of the blade wheels. The defect can be clearly seen in the spectrum by the altered amplitudes. Due to the rotor-synchronous scanning the amplitude spectrum yields discrete values for multiples of the rotor frequency (number of revolutions). This will be discussed in more detail at the end of this section.

In practice the time functions are often scanned with a computer. The discrete Fourier transform is then used to obtain the spectrum. For the individual scanned values of a time function it provides the corresponding values of the Fourier transforms in the frequency range.

The time function x (t) scanned with N values in the time interval T is associated with a Fourier transform (8.17).

$$X(f) = \sum_{K=0}^{N-1} x (kT)_e^{-j2\pi fkT} \quad n = 0,1,\ldots N{-}1 \qquad (8.15)$$

with discrete frequencies

$$f = \frac{n}{N} \cdot \frac{1}{T} \quad n = 0,1,\ldots N{-}1 \qquad (8.16)$$

According to equation (8.15) the calculation of the Fourier transform for a single frequency requires the N-fold calculation of the given function and the summation of these values.

The complex function X (f) is most conveniently determined using the Euler relationship

$$_e{-j\alpha} = \cos \alpha - j \sin \alpha \qquad (8.17)$$

with separation into real and imaginary parts.

For practical applications use is made of a special algorithm for fast calculation of the discrete Fourier transformation. This method is known as the fast Fourier transform (FFT) (8.17, 8.18).

Fig. 8.12 shows the individual amplitudes not in terms of the frequency directly as is customary but in terms of multiples of the machine's rotational speed. It is particularly advantageous when investigating processes with rotating parts to adopt the rotational frequency as a reference variable for the signal frequency.

This is also made use of in the order analysis method (8.19). In this case the frequency analysis is not made in terms of vibrations per time unit as is customary but in terms of vibrations per revolution. A shaft which is out of balance has vibrational frequencies which are always a function of the rotational speed. With

70

increasing speed the frequency spectrum shifts to higher frequencies which are
multiples of the rotational speed. These multiples are designated as orders relative
to the rotational speed in question. The m-th order means that the subfrequency of
a process is m times larger than the frequency of the reference speed. The order

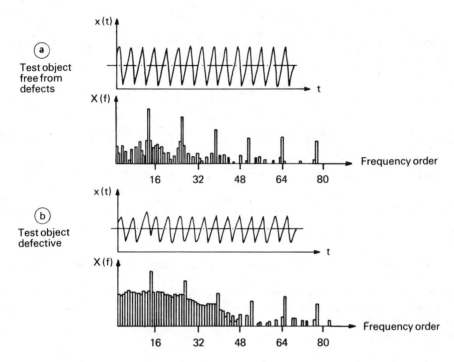

Fig. 8.12   Example of the amplitude spectrum (8.13).

analysis of a process involves recording the amplitude of the individual orders as a
function of the rotational speed. If for instance the first order is examined then from
the signal for each speed the amplitude of the frequency component in Hz has to be
determined which corresponds in magnitude to the number of revolutions per
minute. The same applies to the m-th order for the m-fold frequency component
with respect to the rotational speed.

8.2.1.7   Power density spectrum
Another important signal characteristic is the distribution of the total power of a
process over the individual frequency components. The power per frequency
interval $d\omega$ is called the power density $S(\omega)$. The integral over the power density
then gives the total power (8.20).

$$\overline{x^2(t)} = \int_{0}^{\infty} S(\omega)\, d\omega$$

(8.18)

It makes sense to derive the spectral density analytically from the auto-
correlation function which according to equation (8.10) already has the dimension
of effective power.
   With the help of the Fourier transform, i.e. the transformation from the time

71

domain to the frequency domain the spectral power density $S(\omega)$ of the signal being analysed $x(t)$ is obtained from the correlation function.

The Fourier transform of the auto-correlation function $k_{11}(\tau)$ gives the spectral (auto)–power density $S_{11}(\omega)$ (power spectrum).

$$S_{11}(\omega) = \int_{-\infty}^{\infty} k_{11}(\tau) \cdot e^{-j\omega t}\, d\tau. \tag{8.19}$$

The Fourier transform of the cross-correlation function $k_{12}(\tau)$ gives the spectral cross power density $S_{12}(\omega)$ (cross power spectrum).

$$S_{12}(\omega) = \int_{-\infty}^{\infty} k_{12}(\tau) \cdot e^{-j\omega t}\, d\tau. \tag{8.20}$$

This functional relationship between the spectral power density and the correlation function is also known as the Wiener-Khintchine theorem (8.5, 8.6, 8.20).

For a given cross or auto power spectrum one obtains the corresponding correlation function by the inverse Fourier transform according to equation 8.14.

$$k(\tau) = F^{-1}\{S(\omega)\} = \frac{1}{2\pi} \int_{-\infty}^{+\infty} S(\omega) \cdot e^{+j\omega\tau}\, d\omega \tag{8.21}$$

The unit of the spectral power density is that of power per Hertz bandwidth.

When determining the power density spectrum by actual measurement the signal is filtered for a time duration T by means of a band-pass filter with bandwidth B and mean frequency ω, the result squared and this then averaged and divided by the time duration T and the bandwidth B. This measuring procedure is based on the relationship (8.20, 8.21).

$$S(\omega) = \lim_{B \to 0} \lim_{T \to \infty} \frac{1}{B} \cdot \frac{1}{2T} \int_{-T}^{T} x^2(t,\omega,B)\, dt \tag{8.22}$$

The block diagram for determining the spectral power density is shown in Fig. 8.13. In practice it is preferable to carry out the measurement by means of narrow-band filter sets.

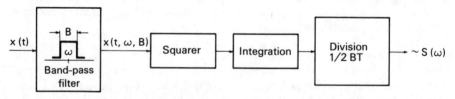

Fig. 8.13 Block diagram for determining the spectral power density.

For practical measurements the observation period for observing the stochastic signals cannot be infinitely long. The following equation cut down to size applies to the error ε resulting from the measurement (8.21).

$$\frac{\varepsilon/\%}{100} \approx \frac{1}{\sqrt{\dfrac{B}{Hz} \cdot \dfrac{T}{s}}} \tag{8.23}$$

If for example the power density spectrum is measured with an error $\varepsilon = 20\%$ and the filter bandwidth $B = 5\,\text{Hz}$ then the signal x (t) must be observed for at least $T = 5\,\text{s}$. Examples of some power density spectra are shown in Fig. 8.14 (8.4).

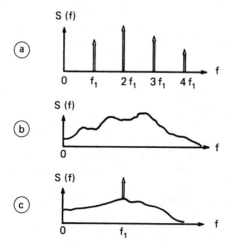

Fig. 8.14   Examples of power density spectra (8.4).

(a) tone with harmonic content
(b) noise
(c) noisy sinusoidal vibration

The power of the individual frequency components can be used as a feature. The power density spectrum in Fig. 8.14b can for example originate from acoustic streaming and friction sounds. The power density spectrum in Fig. 8.14c is obtained for instance with sounds having periodic excitation. In practice despite the contradictory theoretical requirement the power density spectrum is also applied to non-stationary processes taking errors into account. In order to interpret time-varying processes, e.g. the run-up sound or generators, the method of short-term spectral analysis is especially suited for representing the time-frequency-power distribution.

The short-term power density spectrum is measured by restricting the measuring time to a finite, in practice very short time of for instance 50 or 100 ms for audio frequency sounds. In this case successive time segments of the process are analysed. The running together of short-term spectra in a two-dimensional display (running spectra) makes it possible to follow the time changes in the spectral content. This means that dynamic changes are visible in the frequency domain such as impacts or resonance during run-up. The frequency resolution is improved and the harmonic structure is clearer with increasing length of analysis but the ability is lost to follow fast changes in the signal.

8.2.1.8   Homomorphous processing
It is preferable to apply homomorphous signal processing to signals which are concentrated by multiplication or by convolution.

For a linear system with the transfer function L and the two inputs x_1 and x_2 the principle of superposition applies:

$$L(x_1 + x_2) = L(x_1) + L(x_2) \qquad (8.24)$$

The principle of amplification also holds (c is a scalar):

$$L(c \cdot x_1) = c \cdot L(x_1) \qquad (8.25)$$

If apart from addition and multiplication, other operations are allowed such as convolution (equation (8.42)) the principles of superposition and amplification can be generalised (8.1, 8.22, 8.23).

If the operations of addition, multiplication, convolution are designated by the symbols.

    o for the concatenation operation of inputs amongst themselves,
    : for the concatenation operation of the inputs with a scalar,
    ϕ for the concatenation operation of the outputs amongst themselves
    / for the concatenation of the outputs with a scalar,

then the equations (8.24) and (8.25) can be generalised for a system with the transfer function T in the following manner:

$$T(x_1 \, o \, x_2) = T(x_1) \, \phi \, T(x_2) \qquad (8.26)$$

$$T(c : x_1) = c/T(x_1) \qquad (8.27)$$

A system with these properties is called a homomorphous system. The linear system of equations (8.24) and (8.25) is obtained when the operations o and ϕ represent additions and the operations : and / represent multiplications.

If equations (8.26) and (8.27) apply then the transfer system T can be represented by the three subsystems T_o, L, T_o^{-1}. T_o is the so-called characteristic system. The function of this system is to transform the inputs x_1 and x_2 concatenated by the operation o (addition, multiplication, convolution, etc.) in such a way that the permitted concatenations in equations (8.24) and (8.25) become addition and multiplication of a linear system:

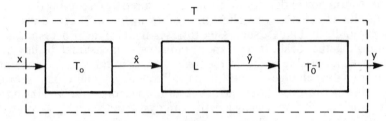

Fig. 8.15   Structure of a homomorphous system (8.22).

$$T_o(x_1 \, o \, x_2) = T_o(x_1) + T_o(x_2) = \hat{x}_1 + \hat{x}_2 \qquad (8.28)$$

$$T_o(c : x_1) = c \cdot T(x_1) = c \cdot \hat{x}_1 \qquad (8.29)$$

74

System L is a linear system which satisfies equations (8.24) and (8.25). The system T_o^{-1} executes the inverse operation to T_o: the operation T_o^{-1} and T_o cancel each other out exactly. For the inverse system T_o^{-1} in Fig. 8.15 the following applies with regard to addition and multiplication:

$$T_o^{-1}(\hat{y}_1 + \hat{y}_2) = T_o^{-1}(\hat{y}_1) \, \phi \, T_o^{-1}(\hat{y}_1) = y_1 \, \phi \, y_2 \qquad (8.30)$$

$$T_o^{-1}(c \cdot \hat{y}_1) = c/T_o^{-1}(\hat{y}_1) = c/y_1 . \qquad (8.31)$$

For the purposes of illustration the application of this method to signals concatenated by multiplication or convolution will be discussed using two examples from picture processing and acoustic processing.

Because of the greater practical importance of the method in acoustics at the present time homomorphous processing comes under the heading of acoustic procedures. As mentioned at the beginning on the one hand the examples in no way indicate that the procedures are exclusive to either acoustics or optics and on the other hand they show how varied the applications can be within these areas.

The first example concerns the processing of multiplied and convolved picture signals (8.22, 8.23).

A scene illuminated with external illumination reflects this light and so generates the impression of local brightness modulation perceived by the observer. This f (x,y) is the multiplication of the intensity of the illumination (index I) with the reflection properties (index R) of the image:

$$f(x,y) = f_I(x,y) \cdot F_R(x,y) \qquad (8.32)$$

A well-defined class of homomorphous systems suggests itself for processing such variables concatenated by multiplication. The characteristic system T_O must possess the properties that

$$T_o(f_1 \cdot f_2) = T_o(f_1) + T_o(f_2) \qquad (8.33)$$

$$T_o(f^\gamma) = \gamma \cdot T_o(f) \qquad (8.34)$$

One function which possesses these properties is the logarithmic function with the exponential function as the associated inverse function T_o^{-1}.

In the following example such a homorphous processing will be used to improve the contrast of a picture (8.22, 8.23).

Very often pictures are spoilt by the large dynamic range of the brightness which varies from the lightest spot on the picture to the darkest shadows. It is desirable to reduce the dynamic range to be processed and still at the same time to improve the recognition of detail. These two opposite requirements can only be satisfied if the illumination and reflection components are influenced separately.

The large dynamic range is mainly produced by the illumination whereas the recognition of detail is essentially determined by the reflection properties of the object. As a rule the illumination component is in a much lower frequency range than the reflection component.

When the components are concatenated by multiplication with one varying faster than the other it is possible by means of homomorphous signal processing to influence the individual components separately. Fig. 8.16 shows the homo-

75

morphous system (equations (8.33) and (8.34)) suitable for reducing the dynamic range with concurrent improvement of detail recognition.

The following applies to the input system:

$$\hat{f} = \log f = \log (f_I . f_R) = \log f_I + \log f_R = \hat{f}_I + \hat{f}_R \tag{8.35}$$

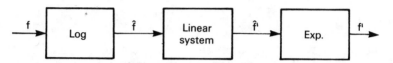

Fig. 8.16 *Homomorphous image processing system for dynamic reduction and contrast enhancement.*

The linear system is an ideal amplifier.

$$\hat{f}^l = \gamma. \hat{f} \tag{8.36}$$

In addition the following applies to the outputs:

$$f^l = \exp \hat{f}^l = \exp (\hat{f}_I^l + \hat{f}_R^l) = \exp \hat{f}_I^l . \exp \hat{f}_R^l = f_I . f_R \tag{8.37}$$

This gives the following relationships between inputs and outputs:

$$f^l = \exp (\gamma. \log f) = f^\gamma = f_I^\gamma . f_R^\gamma \tag{8.38}$$

A dynamic reduction is achieved by an amplification factor $\gamma < 1$ whereas a contrast enhancement is achieved by an amplification factor $\gamma < 1$. This is possible by means of a frequency-dependent amplification (Fig. 8.17) due to the different frequency ranges of the illumination and reflection:

$$f^l = f_I{}^{\gamma_I} . f_R{}^{\gamma_R} \tag{8.39}$$

For example the amplification factor $^{\gamma_I}=0.5$ and $^{\gamma_R}=2$ can be chosen. The outputs can then be obtained from:

$$f^l = f_I^{0.5} . f_R^2 \tag{8.40}$$

This process considerably improves the details of the image in contrast (8.22, 8.23).

The second example of homomorphous processing comes from acoustics. The application of this method to signals concatenated by convolution leads to the so-called cepstrum technique.

With a linear system the Fourier transform $X_o(\omega)$ of the output $x_o(t)$ is obtained by multiplying the Fourier transform $X_i(\omega)$ of the input $x_i(t)$ with the frequency response $G(\omega)$ of the system:

$$X_o(\omega) = G(\omega). X_i(\omega) \tag{8.41}$$

This multiplication in the frequency domain corresponds to a convolution operation in the time domain.

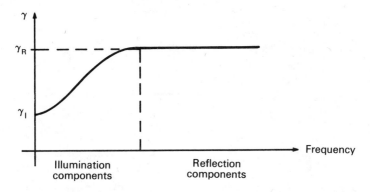

Fig. 8.17 *Frequency response of the amplification for homomorphous image processing (8.22).*

$$x_o(t) = \int_O^t g(t-\tau) \, x_i(\tau) \, d\tau \qquad (8.42)$$

where $g(t)$ is the pulse response of the system also known as the weight function. The frequency response is obtained from the Fourier transform of the weight function:

$$G(\omega) = F\{g(t)\} \qquad (8.43)$$

The cepstrum analysis now makes it possible to separate the effects of the inputs on the one hand and the effects of the system variables $G(\omega)$ on the other hand on the spectral function $X_o(\omega)$ appearing at the system output.

The cepstrum C is understood as the square of the inverse Fourier transform of the logarithmic power spectrum (8.24).

$$C = \left[F^{-1}\{\log | X_o(\omega)|^2\}\right]^2 \qquad (8.44)$$

The cepstrum is obtained by squaring equation (8.41), then taking the logarithm, forming the inverse Fourier transform and finally squaring.

Taking the logarithm pre-emphasises weakly defined spectral lines. Squaring further emphasises the two separated spectral functions. The logarithmic power density spectrum contained in the cepstrum has information concatenated by addition on the transmission system which can be represented by two different spectral periodicities (Fig. 8.18). In order to separate these periodicities from each other frequency response of the logarithmic power density spectrum is subjected to a 'frequency analysis' just as if a time function were being analysed. The function variable of the cepstrum therefore has the dimension of time and has been given the name 'quefrency'. The quefrency expresses the spectral periodicity per Hertz.

Fig. 8.18 shows an example of the application of the cepstrum (8.24). The logarithmic density spectrum of a spoken vowel as a function of the frequency is shown in Fig. 8.18b. The continuous line characterised by periodic minima describes the vibrations of the vocal chords during vowel formation while the dotted curve represents the vocal tract Fig. 8.18c shows the associated cepstrum in which the two components are clearly separated from each other. The sharply defined peak at T expresses the period of the fundamental vibration whereas the

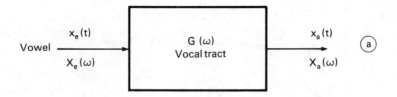

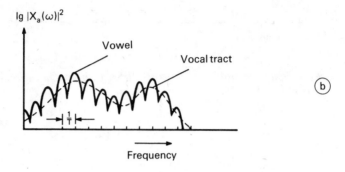

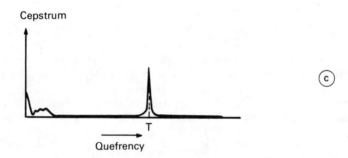

Fig. 8.18   Example of the application of the cepstrum (8.24).

    (a) Transmission system
    (b) Logarithmic power spectrum of a spoken vowel
    (c) Cepstrum

relatively broad curve section stems from very low frequency values of the vocal tract.

    This example shows what the condition is for a separation of the input and system variables by the cepstrum: A broad-band system must be excited by a periodic narrow-band signal.

    The procedure in this example taken from the field of speech recognition can be transferred to industrial systems without any modification. Examples can be found in the quality control of percussion drills (8.25) and in the determination of the timing sequence of internal combustion engines (8.26).

8.2.2.   Image processing
Special processing methods for video signals will now be described. The selection

of the methods is based on their practical application in industry.

The analytical description of individual picture features in the following section 8.2.2.1 (cf. section 4.5.2) assumes that the objects clearly stand out from the background. However this is in general only the case for binary pictures. Therefore apart from a description of the features the question of finding these features is of great importance. This task is made considerably difficult by the presence of several objects in the picture field or the lack of clear separation of the objects from an irrelevant background. Sections 8.2.2.2-8.2.2.4 are therefore specifically devoted to methods for separating and defining desired objects.

8.2.2.1 Analytical description of picture features.

The binary picture shown in Fig. 8.19 B (x,y) can be described as follows (8.27):

$$B(x,y) = \begin{cases} 1 \text{ if } (x,y) \text{ lie inside the silhouette} \\ o \text{ otherwise} \end{cases}$$

An important feature of such a binary picture is the area. It is given by the relationship.

$$F = \int_x \int_y B(x,y) \, dx \, dy \qquad (8.46)$$

The following considerations apply to a scanned picture as often occurs in practice. A picture scanned by a television camera in automatically divided into single lines. Another division system can be generated within each line appropriate counting pulses (Section 4.5.2). If M is the number of picture points within a line (x direction) and N is the number of picture points in the corresponding vertical y direction then the number of picture points lying inside the silhouette is given by

$$A = \sum_{y=1}^{N} \sum_{x=1}^{M} B(x,y) \qquad (8.47)$$

The following then applies to the area (8.27).

$$F = A \cdot \Delta x \cdot \Delta y \qquad (8.48)$$

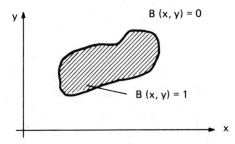

Fig. 8.19   Description of a binary picture.

Here Δx is the division spacing along a line (x direction) and Δy the raster spacing in the y direction (line spacing).

79

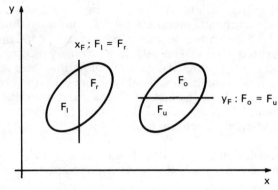

Fig. 8.20   Centre of gravity of the area.

The centre of gravity of the area is another significant feature and is especially important as a reference point in positioning tasks.

According to Fig. 8.20 the centre of gravity of the area is the point of intersection of two lines parallel to the axes and perpendicular to one another which divide the object into two sections with equal area. The two centre of gravity coordinates are defined by (8.27).

$$X_F = \frac{1}{F} \sum_{y=1}^{N} \sum_{x=1}^{M} x \cdot B(x,y) \qquad (8.49)$$

and

$$y_F = \frac{1}{F} \sum_{y=1}^{N} \sum_{x=1}^{M} y \cdot B(x,y) \qquad (8.50)$$

If this general relationship is now adapted to the special conditions of the line scanning of a picture with a television camera then the following relationship is obtained for the centre of gravity with the symbols shown in Fig. 8.21 (8.28):

$$X_F = \frac{r}{2} \cdot \frac{\sum\limits_{n=N_1}^{N2} \sum\limits_{m=1}^{M} (x^2_{n,m2} - x^2_{n,m1})}{\sum\limits_{n=N_1}^{N2} \sum\limits_{m=1}^{M} (x_{n,m2} - x_{n,m1})} \qquad (8.51)$$

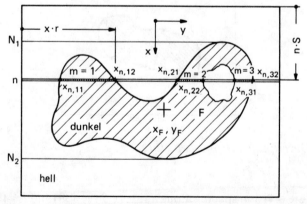

Fig. 8.21   Determination of the centre of gravity x_F, y_F in the television grid (8.28).

80

$$y_F = s \, \frac{\sum\limits_{n=N_1}^{N2} \sum\limits_{m=1}^{M} (x_{n,m2} - x_{n,m1}) \cdot n}{\sum\limits_{n=N_1}^{N2} \sum\limits_{m=1}^{M} (x_{n,m2} - x_{n,m1}) \cdot n} \tag{8.52}$$

where

N_1, N_2 are the first and last scanning lines of the object
x_n, m_1 is the coordinate of the light-dark transitions
x_n, m_2 is the coordinate of the dark-light transition
n is the line number
m is the number of dark segments in the line
s is the line spacing
r is the gap between counting pulses in the line (Fig. 4.22)

The implementation of such a system is considered in Section 10.1.2.

Another important feature is the principal moments of inertia I_1 and I_2. These variables commonly used in mechanics are derived from the axial moment of inertia of the area and the centrifugal moment (8.29). The axial geometrical moment of inertia with respect to an axis in the plane of the area is the sum of the products of the area elements and the square of their vertical distances from the axis.

The centrifugal moment is referred to two axes lying in the plane and is the sum of the products of the area elements and the products of their distances from both axes.

The geometrical axial moments of inertia I_1 and I_y with respect to the x and y axes can be defined by the following relationship in a suitable form for picture processing (8.27):

$$I_x = \frac{1}{F} \sum_{y=1}^{N} \sum_{x=1}^{M} (x-x_F)^2 \cdot B(x,y) \tag{8.53}$$

$$I_y = \frac{1}{F} \sum_{y=1}^{N} \sum_{x=1}^{M} (y-y_F)^2 \cdot B(x,y) \tag{8.54}$$

The centrifugal moment is defined by:

$$I_{xy} = \frac{1}{F} \sum_{y=1}^{N} \sum_{x=1}^{M} (x-x_F) \cdot (y-y_F) \cdot B(x,y) \tag{8.55}$$

The principal moments of inertia can then be obtained from the relationship (8.29).

$$I_{\frac{1}{2}} = \frac{1}{2}(I_x + I_y) \pm \sqrt{\tfrac{1}{4}(I_x - I_y)^2 + I_{xy}^2} \tag{8.56}$$

These principal moments of inertia are the smallest and largest geometrical moment of inertia with respect to two mutually perpendicular axes – the principal axes.

The dimension of the moment of inertia is the fourth power of a unit of length.

The principal moments of inertia are especially suitable as features because they have rotational invariance.

A practical example of using these features is discussed in section 10.1.7.

Another feature which is independent of the rotational position is the roundness of a body which is defined as (8.30).

$$R = \frac{1}{F^2} (I_x + I_y)$$

(8.57)

This is a measure of how much the body is related to a circle for which with

$$I_x = I_y = \frac{r^4 . \pi}{4} \text{ the value } R = \frac{1}{2\pi} \text{ applies.}$$

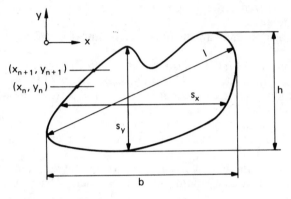

Fig. 8.22   Features in a binary picture.

Other features commonly used are (Fig. 8.22).
– the maximum dimension of the object in the x and y directions:

$$b = \max \{\Delta x\}$$

(8.58)

and

$$h = \max \{\Delta y\} = | N_2 - N_1 |$$

(8.59)

where Δx and Δy are the differences of the coordinates of all points of intersection (x_n, y_n) of the n scan lines $N_1 < n < N_2$ (Fig. 8.21) with the external contour of the object;
– the maximum chord s_x and s_y in the x and y directions:

$$s_x = \max \{\Delta x\}_{y=\text{const}}$$

(8.60)

and

$$s_y = \max \{\Delta y\}_{x=\text{const}}$$

(8.61)

– the longest chord

$$1 = \max \{ \sqrt{x_n - x_n')^2 + (y_n - y_n')^2} \}$$

(8.62)

$$\text{with } N_1 < (n,n') < N_2$$

82

At the same time this value specifies the diameter of the smallest circumscribing circle;

– the diameter of a circle with the same area as the pattern

$$d = 2\sqrt{F/\pi} \qquad (8.63)$$

– the perimeter of the external contour

$$U = \sum_n \sqrt{(x_{n+1} - x_n)^2 + (y_{n+1} - y_n)^2} \qquad (8.64)$$

where the distances between neighbouring scanned contour intersection points are added round the entire external contour;

– a measure of extension

$$a = l^2/F \qquad (8.65)$$

– a measure of compactness

$$b = F/b.h \qquad (8.66)$$

which relates the area of the pattern to the area of the smallest circumscribing rectangle;

– the number of holes in the object (cf. Section 8.2.2.4);
– the area of the holes in the object.

8.2.2.2 Template matching

An important processing method for patterns $f(x,y)$ involves matching the picture to a known object by means of a template. This makes it possible to find a certain object in the presence of several objects or in distorted pictures.

The triangle shown in Fig. 8.23a can be detected by moving a template (Fig. 2.23b) until it is completely light inside on overlapping. However with this simple mask the rectangle shown in the diagram would also make the template light inside on overlapping.

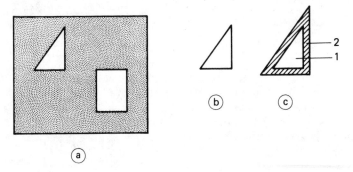

Fig. 8.23   Example of template matching.

      (a) Picture field with patterns
      (b) Simple template
      (c) Improved template

This can be prevented by using an improved template in which the light triangle is surrounded by a second area (Fig. 8.23c). The required triangle will always be present if the inside (1) is light and at the same time the outside (2) is dark.

The determination of the degree of lightness or darkness in the defined picture area can be analog as well as digital. An example of analog determination is the measurement of the light incident on the different areas by means of photodetectors. In the digital case the picture is transformed into a binary image in which for example '1' is assigned to the light areas and '0' to the dark areas. The number of ones covered by the template in the individual areas is then a measure for the matching where in the present example (the template of Fig. 8.23c) the inside area should present as many ones as possible and the outside area as few ones as possible.

The following considerations for determining a measure of the matching between template and picture are based on a binary picture divided into individual picture elements. Such a picture grid with the brightness values B (i,j) of the individual picture elements is shown in Fig. 8.24. In the general case an I x J matrix is produced ($i=1,\ldots,I$ and $j=1,\ldots,J$).

In the same way a mask M(p,q) can be defined as a P x Q matrix with a picture field D ($p=1,\ldots,P$; $q=1,\ldots,Q$). This mask covers the points B $(i-p, j-q)$ in the picture field B (i,j).

A possible measure for the matching between mask and picture is the sum of all the brightness differences between picture and mask (8.31).

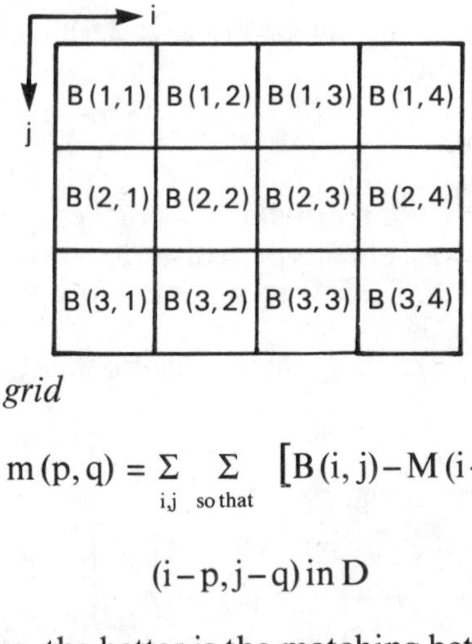

Fig. 8.24   Picture field grid

$$m(p,q) = \sum_{i,j} \sum_{\text{so that}} [B(i,j) - M(i-p, j-q)] \qquad (8.67)$$

$$(i-p, j-q) \text{ in } D$$

The smaller this value, the better is the matching between mask and picture. A considerable disadvantage of this measure however is that even when there is no matching the value m(p,q) can be very small. This is the case when $B > M$ for some of the terms and $M > B$ for other terms. Because of the different signs the results cancel each other out for the wrong reason.

This disadvantage is avoided by the following measure (8.31).

$$m'(p,q) = \left\{ \sum_i \sum_j [B(i,j) - M(i-p, j-q)]^2 \right\}^{1/2} \qquad (8.68)$$

Squaring this gives

$$m'^2(p,q) = \sum_i \sum_j [B^2(i,j) - 2B(i,j) \cdot M(i-p, j-q) + M^2(i-p, j-q)] \qquad (8.69)$$

84

The last term of this equation is exclusively determined by the geometry of the mask (mask-specific constant).

The first term provides a measure of the light energy of the picture and is purely a function of the picture.

The middle term is identical to the cross-correlation function k_{BM} (p,q) of the two functions B and M.

$$k_{BM}(p,q) = \sum_i \sum_j B(i,j) . M(i-p, j-q) \tag{8.70}$$

This expression is used as a definition for the matching of mask and picture. The greatest matching is then achieved when k_{BM} has a maximum value.

It is especially simple to carry out the cross-correlation optically (8.32, 8.33). The principle of an incoherent optical correlator is shown in Fig. 8.25. The required cross-correlation function k_{BM} between object and reference mask is obtained at the output of the photodetector from the following relationship:

$$k_{BM} = \int_{x=0}^{x_0} \int_{y=0}^{y_0} B(x,y) . M(x-p, y-q) \, dx \, dy \tag{8.71}$$

where x_0 is the picture width and y_0 the picture height. An example for a sensor based on optical correlation is presented in section 10.1.1.

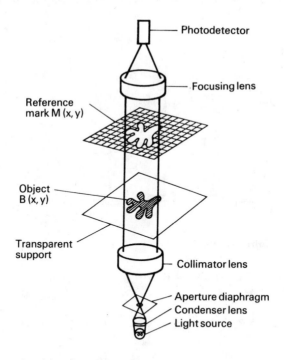

Fig. 8.25   Principle of an optical correlator.

8.2.2.3   Determination of contour lines from grey-tone changes

In many industrial applications it is impossible to generate a well-defined binary picture so that grey-tone picture processing is unavoidable. An example is the recognition of workpieces on pallets where it is impossible to prevent the background of the workpiece from becoming dirty so that the picture becomes ill-defined.

One method of processing grey-tone pictures is to transform them into line drawings whereby the contrast boundaries of the picture are replaced by lines (Fig. 4.16c).

It is possible to determine the points at which changes in the grey-tone occur by differentiating the pattern $f(x,y)$. In order to find the contour line it is necessary to find the magnitude and direction of the change in grey-tone. In this connection it is convenient to calculate the gradient. Its magnitude specifies the maximum change in the brightness at a picture point; its direction is that of the largest change in the brightness.

The gradient in the Cartesian system of coordinates is defined by the relationship

$$g(x,y) = \frac{\partial f(x,y)}{\partial x} e_x + \frac{\partial f(x,y)}{\partial y} e_y \tag{8.72}$$

where $f(x,y)$ is the grey scale value of a pattern at the point (x,y) and e_x and e_y are the unit vectors in the direction of the axes.

Fig. 8.26 shows a point $P = (x_i, y_j)$ which is surrounded by eight neighbouring points A–H (grid). The coordinates of these points are specified on the corresponding axes.

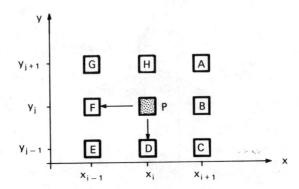

Fig. 8.26   Determination of the gradient.

With a grid picture of this type the following operation can be conveniently used with a digital computer (8.1).

$$g = \frac{f(x_i, y_j) - f(x_{i-1}, y_j)}{\Delta x} e_x + \frac{f(x_i, y_j) - f(x_i y_{j-1})}{\Delta y} e_y, \tag{8.73}$$

where the distances of neighbouring points (x_{i-1}, y_j) and $x_i, y_{j-1})$ are assumed constant with Δx and Δy.

Since only three points (P, F, D) in an unsymmetrical position are involved in the operation there is a high sensitivity to any distortion. A distortion at only one point results in a considerable change in the magnitude and direction of the gradient.

The suppression of distortion is made more efficient if more neighbouring points to the point in question (x,y) are involved in the calculation of the gradient (8.34). If

8 neighbouring points are used the change in the grey scale value is determined from:

$$g(x,y) = a \cdot e_x + b \cdot e_y, \qquad (8.74)$$

with

$$a = f(x_{i+1}, y_{j+1}) + f(x_{i+1}, y_j) + f(x_{i+1}, y_{j-1}) - f(x_{i-1}, y_{j+1})$$

$$- f(x_{i-1}, y_j) - f(x_{i-1}, y_{j-1}) \qquad (8.75)$$

$$= A + B + C - (G + F + E)$$

$$b = f(x_{i-1}, y_{j-1}) + f(x_i, y_{j-1}) + f(x_{i+1}, y_{j-1}) - f(x_{i-1}, y_{j+1})$$

$$- f(x_i, y_{j+1}) - f(x_{i+1}, y_{j+1}) \qquad (8.76)$$

$$= E + D + C - (G + H + A).$$

The components a and b of the gradient are obtained as can be seen from Fig. 8.26 as differences in brightness between two lines and columns adjacent to the point P respectively.

Other combinations are also common for summarising the brightness values around the picture point in question (8.1). In particular further improvement in the suppression of distortion is obtained by heavier weighting of the picture points in the immediate vicinity of the point being investigated P (x_i, y_j). In this case the components of the gradient vector can for example be determined as follows:

$$a = A + 2B + C - (G + 2F + E) \qquad (8.77)$$

$$b = E + 2D + C - (G + 2H + A). \qquad (8.78)$$

The magnitude of the gradient for the point P is obtained from the relationship

$$|g| = \sqrt{a^2 + b^2} \; . \qquad (8.79)$$

In order to find the contour, points with changes in brightness above a threshold S are given the value 1 and below this threshold the value O:

$$B(x,y) = \begin{array}{l} 1, \text{ if } |g| > S \\ 0, \text{ if } |g| < S \end{array} \qquad (8.80)$$

Fig. 8.27 gives a simple example showing how a contour picture can be generated in this way from a grey-tone picture.

In practice a value is often chosen for the threshold S which is proportional to the average brightness of the picture (8.34).

$$S = \text{const} \int_x \int_y f(x,y) \, dxdy \qquad (8.81)$$

This procedure will be illustrated by an example in section 10.1.13.

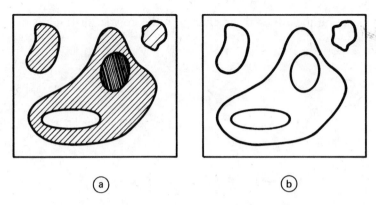

Fig. 8.27 Generation of a contour picture (b) from changes in brightness in the grey-tone picture.

8.2.2.4 Separation of objects

In many applications it cannot be assumed that there is only one object to be recognised in the picture field. Whenever this is possible efforts will be made in industry to isolate the objects for the purposes of simplification. However in some applications this is to costly or leads to impermissibly long cycle times in the manufacturing process.

A possible method of object separation (segmentation) is now presented. Fig. 8.28 shows a scene with several objects. The picture field involves two complete workpieces as well as parts of two other workpieces of the same type. On the one hand there is the difficulty of assigning the structures in the objects to individual workpieces during picture processing. On the other hand the objects have to be discriminated in terms of feature extraction, orientation and classification. The algorithm for object separation (8.35, 8.36) presented here as an example is adapted to the line scanning of a picture which is standard for image transducers. Since only two successive lines are examined at any time the memory requirements are small. A fast electronic circuit therefore suggests itself for on-line processing synchronised with the picture line scanning.

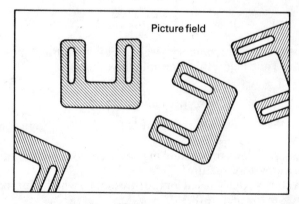

Fig. 8.28 Picture field with several workpieces.

When line scanning a picture field for every two successive lines the x co-ordinates of the point of intersection of the lines with the object are stored together with a characteristic for the object membership defined below.

The determination of the object membership as well as the possible object states in the two lines such as continuation, new formation, end and joining will be illustrated distinguishing different cases.

Fig 8.29 shows the contour points on two successive lines k-1 and k. The case is shown where the object is continued between the lines k-1 and 1, i.e there is no change in the number of objects.

The indices i and j do not always have to be the same. For instance differences can arise as the result of new objects or objects ending in one of the lines.

The following simple plausibility conditions can be established for the continuation of an unchanged object (8.35):

$$x_{j-1} < x_i < x_{j+1} , \tag{8.82}$$

$$x_{i-1} < x_j < x_{i+1} , \tag{8.83}$$

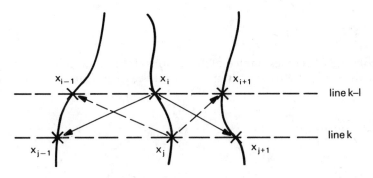

Fig. 8.29   Unchanged number of objects.

$$l_i = l_j , \tag{8.84}$$

$$r_i + r_j . \tag{8.85}$$

Equations (8.82) and (8.83) describe the simple fact that for a continuation the coordinate value of the object contour always lies between the coordinate values of the left and right neighbours by arrows in Fig. 8.29 (a prerequisite for this is a sufficiently narrow line spacing).

Equations (8.84) and (8.85) mean that the number of left and right contour points has remained the same in both lines and that the left and right contour points are also continued in the new lines as left and right contour points respectively. All contour points are assigned to the same object with the characteristic $O(k-1) = O(k)$.

Every time an object arrives or disappears one of the conditions (8.82) – (8.85) will be violated.

If a new object is discovered in a scan line then the situations shown in Fig. 8.30 are possible. A hole in a workpiece is also considered to be a new object. If during line scanning a new object is encountered there will be more contour points than in the previous line. A contour point i is not continued as a contour point j. The following condition can thus be derived for a new object (8.35):

$$x_i > x_{j+1} . \tag{8.86}$$

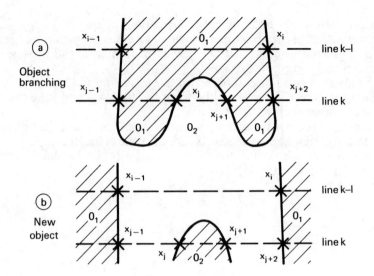

Fig. 8.30   Object genesis

This condition becomes immediately obvious if the contour intersection points of the two lines (Fig. 8.30) are written underneath each other in the stored sequence:

Line k-1   : x_{i-1} x_i
Object assignment : O_1 O_1
Line k x_{j-1} x_j x_{j+1} x_j+2
Object assignment O_1 O_2 O_2 O_1

Criterion 8.82 is violated and equation 8.86 is satisfied. At the same time the number of contour points in the new line has increased.

The two newly inserted points in line k are assigned to the next object number (in this case O_2).

Fig. 8.31a shows the joining of two initially separated objects in Fig. 8.31b the ending of an object.

The condition for these two cases is as follows (8.35):

$$x_j > x_{i+1} . \tag{8.87}$$

In this case the following values are stored:

Line k-1   : x_{i-1} x_i x_{i-1} x_{i+2}
Object assignment : O_{xi-1} O_{x_i} $O_{x_{i+1}}$ $O_{x_{i+2}}$
Line k   : x_{j-1} x_j

Criterion (8.83) is violated x_j does not have its coordinates between the two neighbours of the previous line marked by arrows. However according to the coordinate values of Fig. 8.31 equation (8.87) is satisfied. At the same time the number of contour points in the new line has decreased.

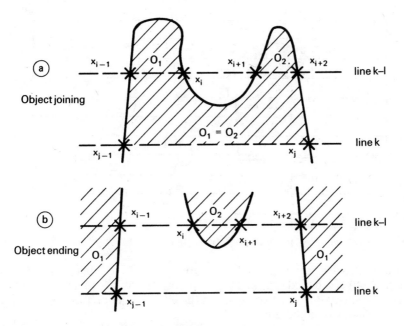

(a) x_{i-1} O_1 x_{i+1} O_2 x_{i+2} line k–l

Object joining x_i

$O_1 = O_2$ line k

x_{j-1} x_j

(b) x_{i-1} O_2 x_{i+2} line k–l

Object ending O_1 x_i x_{i+1} O_1 line k

x_{j-1} x_j

Fig. 8.31   Joining and ending of objects.

The discrimination between the two possible cases of object union or object termination allowed by equation (8.87) is made on the basis of the assigned object numbers of the previous line k-1.

When two objects come together the contour points x_i and x_{i+1} have different object numbers according to Fig. 8.21a (8.35).

$$O_{xj}(k\text{-}l) \neq O_{x_{i+1}}(k\text{-}l) \ . \tag{8.88}$$

In this case the contour point x_i lies above the point of line k for which the condition (8.87) is satisfied.

When an object is terminated these contour points have the same object number (Fig. 8.21b):

$$O_{x_i}(k\text{-}l) = O_{xj+1}(k\text{-}l) \ . \tag{8.89}$$

The following example will further illustrate the methodology and problems of object separation.

Fig. 8.32 shows a picture field B with two identical objects to be separated as well as the associated contour points. Table 8.1 gives a summary of the results for the 10 scan lines.

The picture field is scanned over 10 lines. The two objects to be separated I and II consist of the subobjects I = 1 + 2 and II = 3 + 4 respectively. Since conditions (8.82) – (8.89) have been derived for objects lying inside an area defined by an outer perimeter the left and right edge of the picture field (coordinates 0 and 140) will be also taken into consideration here.

91

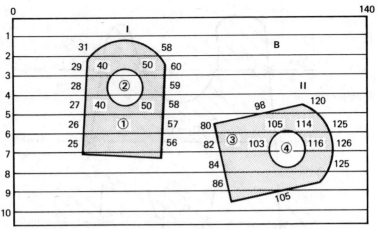

Fig. 8.32   Example of the separation of two objects.

Line 1 only involves the edge of the picture field B. Line 2 takes in subobject ɪ. Condition (8.86) is satisfied (marked by the arrow in Table 8.1): a new object 1 has been introduced between the two points at the edges of the picture field. A similar process is involved for object 2 in line 3.

The continuation conditions (8.82) and (8.83) are satisfied between lines 3 and 4.

In line 5 the continuation condition is violated and condition (8.87) is satisfied for the contour point with coordinate 58. This means either a union or the termination of an object. According to equation (8.89) $O_{x_i}(k-1) = O_{x_{i+1}}(k-1)$: in the previous line $k-1 = 4$ the object assignments are the same (object 2) for the point (x_i) and its righthand neighbour $(x_i + 1)$ lying above the contour point with coordinate 58. A termination of the subobject 2 is therefore involved. Since subobject 2 has come from object 1 (line 3) and subobject 1 has not yet finished the resulting object receives the old object number 1.

In contrast to line 4 the case occurs in line 5 that the number of contour points remains the same although an object (here subobject 2) terminates. The satisfying of the condition for terminating an object with the number of contour points remaining the same means that a new object has entered the scene. This object is intoduced in line 5 with the new object number 3.

In line 6 the condition for a new object (4) is satisfied.

The continuation condition is satisfied in lines 6 and 7.

In line 8 the condition (8.87) for termination is satisfied for object 1. At the same time condition (8.89) is satisfied. This condition applies to the case shown in Fig. 8.31b. According to this object 1 which includes subobject 2 is a self-contained object (I) in the picture field.

Since out of the 8 contour points of line 7 only 4 are on the next line another subobject must have terminated. In this connection the two contour points (coordinates 84 and 125) remaining within the picture field are compared with those of line 7. This produces the following:

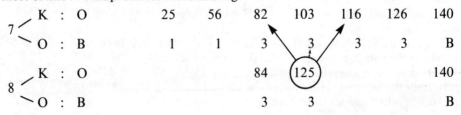

Table 8.1. Contour points (K) and object assignments (O) in lines 1-10 in the example of Fig. 8.32 for object separation.

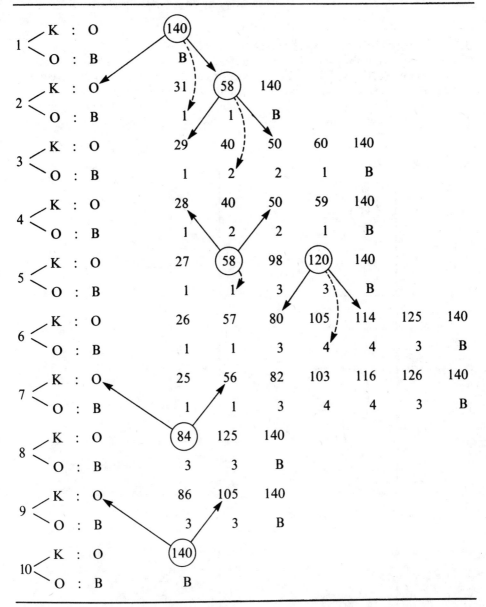

The termination is satisfied for object 4. The terminated object 4 inside subobject 3 therefore becomes subobject 3.

The continuation condition is satisfied between lines 8 and 9.

In line 10 subobject 3 which contains subobject 4 is assessed as another self-contained object (II) in the picture field.

Should the picture field also include sections of objects at the edges as shown in Fig. 8.28 then these can be distinguished by for example measuring the area (Section 8.2.2.1) or by the coincidence of a contour point with the edge of the picture. This example clearly shows the difficulties involved in the apparently easy task of separating two simple objects. It is certainly worth taking simple measures wherever possible to ensure that there is only one object in the picture field.

8.3 Classification methods

Various procedures for preprocessing and feature extraction have been considered in the previous sections. After the preprocessing and feature extraction stage comes the classification stage where a decision is made on the pattern class involved.

This means that the multi-dimensional feature vector c (equation 8.1) generated in the preprocessing stage is assigned a specific class. Corresponding to the

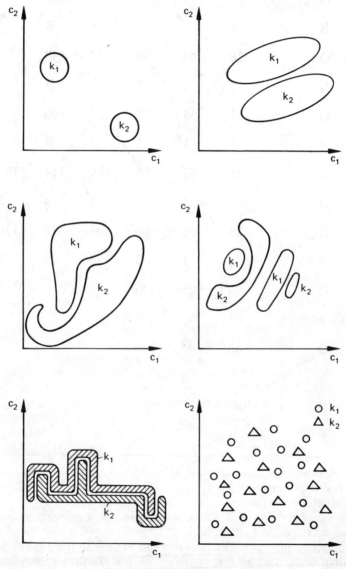

Fig. 8.33 Examples of class clusters in feature space (8.1).

representation of a vector a geometrical approach suggests itself with the feature vector c as a point in an n-dimensional space.

Examples for patterns in 2-dimensional feature space $c = (c_1, c_2)$ are given in Fig. 8.33.

An essential requirement for all automatic classification procedures is that the feature vectors of one class occupy a compact area in the associated multi-dimensional feature space. One speaks here of clusters. The more compact and isolated are the areas shown for example in Fig. 8.33 the simpler and more accurate is the classification. Every effort is made to find features which satisfy these requirements. This task is more complicated going from top to bottom for the cases shown in Fig. 8.33.

Finding and selecting features involves a learning process: using selected features a human (teacher) checks whether an object is assigned to the correct class. The learnt feature vectors are stored as an internal model of the object (Section 4.5.5). During the actual measuring process the same features are obtained from the unknown object and compared with the stored features.

A large number of procedures are given in the pattern recognition literature for solving the classification problem. However only a few of these procedures can be used in practical sensor applications because of the requirements of economic viability and simple solutions. Selected classification methods will now be considered which are especially suitable for industrial sensor tasks.

8.3.1 Linear classifier

Linear classifiers are especially interesting because of their easy implementation.

Two classes are called linearly separable when two linear discrimination functions exist or one dividing plane exists.

With m classes a maximum of m (m-1)/2 dividing surfaces are produced (8.37) which however are not always necessary for complete demarcation of the class areas.

A decision is made that the feature vector c belongs to class k if it lies on the side of the dividing surface assigned to the class k.

Fig. 8.34 shows two pattern classes k_1 and k_2 in the feature space (c_1, c_2) which can be separated by a straight line e (c) $w_0 + w_1 c_1 + w_2 c_2 = 0$ where the parameters w_i determine the position of the line. For every pattern of class $k_1 e(c)$ gives a positive value and for every pattern of class k_2 it gives a negative value. The function e(c) can therefore serve as a discriminating function for an unknown pattern.

If the pattern lies exactly on the separating line, i.e. e(c)=0, then no statement can be made on class membership.

Starting with this simple two-dimensional example the corresponding discriminating function can be derived for the two-class problem in the n-dimensional case (8.2):

$$e(c) = w_0 + w_1 c_1 + w_2 c_2 + \ldots + w_n c_n$$

$$= \sum_{i=1}^{n} w_i c_i + w_0 \ . \tag{8.90}$$

The vector $w = (w_0, w_1 w_2, \ldots, w_n)$ is called the weighting vector. It determines the position of the dividing surface in the $(c_1 \ldots, c_n)$ space. In the 2-dimensional case the discriminating function is a straight line and in the three-dimensional case a plane.

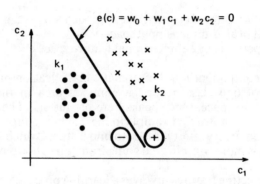

Fig. 8.34   Linear discrimination function for the two-class problem.

The discrimination rule for the two-class problem can be expressed as

$$e(c) = \begin{cases} > O, & \text{if pattern } f(x,y) \text{ in class 1} \\ < O, & \text{if pattern } f(x,y) \text{ in class 2} \end{cases} \qquad (8.91)$$

If this is extended to the multiclass problem with $j = 1,2, \ldots, m$ classes to be discriminated then there will be m discrimination functions of the type

$$e_j(c) = w_{jo} + w_{j1} c_1 + w_{j2} c_2 + \ldots w_{jn} c_n$$

$$= \sum_{i=1}^{n} w_{ji} c_i + w_{jo}, \quad j = 1,2,\ldots,m \qquad (8.92)$$

The discrimination rule for a feature vector to be classified is then

$$e_j(c) = \begin{cases} > O & \text{if pattern } f(x,y) \text{ in class } k_j \\ < O & \text{otherwise} \end{cases} \qquad (8.93)$$

The use of linear discrimination functions will serve as a simple example. Fig. 8.35a shows three classes k_1, k_2, k_3 to be distinguished in the 2 dimensional feature space (c_1, c_2). This is done by the three straight lines $e_1(c) = O$, $e_2(c) = O$ and $e_3(c) = O$. For this case given in Fig. 8.35b the following applies to the discrimination functions.

$$e_1(c) = -c_1 + c_2$$
$$e_2(c) = c_1 + c_2 - 5$$
$$e_3(c) = -c_2 + 1 \qquad ,$$

which yields the following equations for the dividing lines:

$$-c_1 + c_2 = O$$
$$c_1 + c_2 - 5 = O$$
$$-c_2 + 1 = O$$

This gives the areas for the individual pattern classes marked in Fig. 8.35b. If for example a pattern is to be classified with the feature vector $c = (6,5)$ then by

96

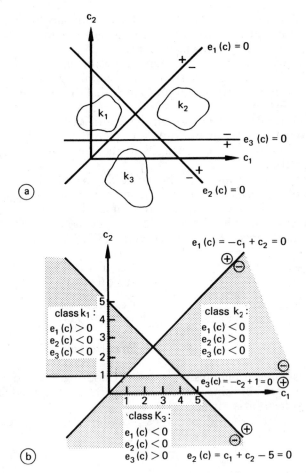

Fig. 8.35   Example of class separation for the linear classifier.

substituting in the discrimination functions we have $e_1(c) = -1. e_2(c) = 6, e_3(c) = -4$. Since $e_2(c) > O$ and $e_1(c) < O$ and $e_3(c) < O$ equation (8.93) gives an assignment to class k_2.

In the multidimensional case the dividing lines cannot be determined on sight. We shall now consider two procedures for determining the components of the weighting vector for this case.

8.3.1.1   Determination of the parameters of the linear classifier

The position of the dividing plane is determined by the weighting vector. An iterative method for determining the components of the weighting vector will now be given (8.2, 8.38). For reasons of clarity the two-class problem will be considered as a special case.

Equation (8.90) for the discrimination function can be modified as follows:

$$e(c) = \sum_{i=o}^{n} w_i c_i \; , \tag{8.94}$$

where c_o is set equal to 1. This means that the dividing surface $e(c) = O$ goes through the origin of the now $(n+1)$-dimensional space.

If the weighting vectors w are initially fixed arbitrarily the correct classification $e(c) > O$ for all feature vectors from class 1 or $e(c) < O$ for all feature vectors from class 2 will certainly only happen randomly.

In the following method the weighting coefficients are altered every time a feature vector is incorrectly classified. Depending on the class membership the weighting vectors are corrected in such a way that the class conditions $e(c) > O$ or $e(c) < O$ are satisfied.

If for example after l correction for the feature vector $c_{(l)}$ of the class 1 with the associated weighting vector $w_{(l)}$ $e(c) < O$ then the classification is wrong. The weighting vector $w_{(l)}$ is then changed to the next value $w_{(l+1)}$. This change is continued until $e(c) > O$.

At this point the meaning of the index will be clarified: w_i denotes the i-th component of the weighting vector; the index in brackets $w_{(i)}$ on the other hand denotes the weighting vector after l correction steps.

The following plausible iteration rule can be devised for the stepwise determination of the components of the weighting vector (8.2, 8.37).

$$w_{(l+1)} = w_{(l)} + a \cdot c_{(l)} \qquad \text{if } c_{(l)} \text{ in class 1} \text{ and } e(c) < 0 \qquad (8.95)$$

$$w_{(l+1)} = w_{(l)} - a \cdot c_{(l)} \qquad \text{if } c_{(l)} \text{ in class 2} \text{ and } e(c) > O$$

$$w_{(l+1)} = w_{(l)} \qquad \text{for correct classification}$$

Here the symbol a denotes a constant. Choosing $a = 1$ does not mean any restriction in generality since in any case incorrectly classified feature is changed until the classification is correct.

A simple example should make this procedure clear. It is required to find the

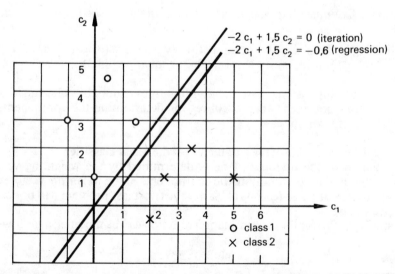

Fig. 8.36 *Example for the determination of the components of the weighting vector.*

98

components of the weighting vector which determines the dividing line for the eight feature vectors of the two classes shown in Fig. 8.36.

The feature vectors of class 1 are:

$$\begin{pmatrix} 0 \\ 1 \end{pmatrix} , \begin{pmatrix} -1 \\ 3 \end{pmatrix} , \begin{pmatrix} 0.5 \\ 4.5 \end{pmatrix} , \begin{pmatrix} 1.5 \\ 3 \end{pmatrix}$$

The feature vectors for class 2 are:

$$\begin{pmatrix} 2 \\ -0.5 \end{pmatrix} , \begin{pmatrix} 2.5 \\ 1 \end{pmatrix} , \begin{pmatrix} 3.5 \\ 2 \end{pmatrix} , \begin{pmatrix} 5 \\ 1 \end{pmatrix}$$

Table 8.2 gives the steps carried out according to the rule (8.95) for determining the weighting vector w. Here k has been arbitrarily set equal to 1 and the starting point chosen as $w = \begin{pmatrix} 0 \\ 0 \\ 0 \end{pmatrix}$. The components of the feature vector are given in the rows on the lefthand side with c_o set equal to 1.

Beginning from the chosen starting point the table is executed line by line according to the specifications of the learning rule (8.95). A correction is only made if $w_{(l+1)} \neq w_{(l)}$. In this case $w_{(l+1)} = w_{(l)}$ is satisfied after the second iteration step and the process terminated. The learning rule produces the dividing line $-2c_1 + 1.5c_2 = O$ (Fig. 8.36).

It can be shown that the learning rule (8.95) always provides a correct classification after a finite number of iteration steps if the classes are at all linearly separable (8.2).

Another method of determining the components of the weighting vector is based on target variables of the type

$$z_j = \begin{cases} a_j \text{ if c in class j} \\ a \neq a_j \text{ otherwise} \end{cases} \tag{8.96}$$

which characterise the class areas. Here a_j and a_j are predefined constants. It is required to find the linear discrimination function $e_j (c)$, which approximates to the corresponding target variable in a class as well as possible, e.g. in the root mean square (linear regression) (8.2, 8.37). It can be shown that the required n components of the weighting vector w_j of the class j for the discrimination function equation (8.92) can be obtained from the following matrix equation (8.2):

$$M \cdot w_j = s_j \cdot \tag{8.97}$$

In this case the coefficients of the matrix M are calculated from the predefined feature vectors as follows:

$$M = \begin{pmatrix} 1 & \bar{c}_2 & \bar{c}_2 & \cdot & \cdot & \cdot & \bar{c}_n \\ \bar{c}_1 & \bar{c}_1 & \bar{c}_1\bar{c}_2 & \cdot & \cdot & \cdot & \bar{c}_1\bar{c}_n \\ \bar{c}_2 & \bar{c}_1\bar{c}_2 & \bar{c}_2^2 & \cdot & \cdot & \cdot & \bar{c}_2\bar{c}_n \\ \cdot & \cdot & \cdot & & & & \cdot \\ \cdot & \cdot & \cdot & & & & \cdot \\ \cdot & \cdot & \cdot & & & & \cdot \\ \bar{c}_n & \bar{c}_1\bar{c}_n & \bar{c}_2\bar{c}_n & & & & \bar{c}_n^2 \end{pmatrix} \tag{8.98}$$

The bar means the formation of the average over all classes. This means that for

Table 8.2 Learning steps for determining the components of the weighting vector using the learning rule (8.95) (example).

	Feature vector			Class	Weighting vector $w_{(l)}$			e(c)	Correction	Weighting vector $w_{(l+1)}$		
	c_0	c_1	c_2		w_0	w_1	w_2			w_0	w_1	w_2
First step	1	0	1	1	0	0	0	0	Yes	1	0	1
	1	−1	3	1	1	0	1	4 (> 0)	No	1	0	1
	1	0,5	4,5	1	1	0	1	5,5(> 0)	No	1	0	1
	1	1,5	3	1	1	0	1	4 (> 0)	No	1	0	1
	1	2	−0,5	2	1	0	1	0,5(> 0)	Yes	0	−2	1,5
	1	2,5	1	2	0	−2	1,5	−3,5(< 0)	No	0	−2	1,5
	1	3,5	2	2	0	−2	1,5	−4 (< 0)	No	0	−2	1,5
	1	5	1	2	0	−2	1,5	−8,5(< 0)	No	0	−2	1,5
Second step	1	0	1	1	0	−2	1,5	1,5(> 0)	No	0	−2	1,5
	1	−1	3	1	0	−2	1,5	6,5(> 0)	No	0	−2	1,5
	1	0,5	4,5	1	0	−2	1,5	5,75(> 0)	No	0	−2	1,5
	1	1,5	3	1	0	−2	1,5	1,5(> 0)	No	0	−2	1,5
	1	2	−0,5	2	0	−2	1,5	−4,75(< 0)	No	0	−2	1,5
	1	2,5	1	2	0	−2	1,5	−3,5(< 0)	No	0	−2	1,5
	1	3,5	2	2	0	−2	1,5	−4 (< 0)	No	0	−2	1,5
	1	5	1	2	0	−2	1,5	−8,5(< 0)	No	0	−2	1,5

instance when determining $\bar{c}_1$ all the first components of all feature vectors of all classes are added and the result divided by the total number of all feature vectors.

The righthand side of equation (8.97) is a vector which describes the dispersion between the feature and the target variable. The following applies to this

$(j = 1, \ldots, m)\,(8.2)$

$$S_j = \begin{pmatrix} \overline{z_j} \\ \overline{z_j c_1} \\ \overline{z_j c_2} \\ \cdot \\ \cdot \\ \cdot \\ \overline{z_j c_n} \end{pmatrix} \qquad (8.99)$$

The average should also be formed over all classes here. The corresponding components of the feature vectors should be multiplied for all feature vectors of all classes with the associated target variables, the individual results summed and divided by the total number of all feature vectors of the classes.

When using these relationships the dividing surface is determined between one class and all other classes in every case. There are n+1 equations to be solved for each class. This means that for m classes there are m (n+1) equations. Appropriate calculation procedures are given in (8.2).

For the purposes of illustration the procedure will be applied to the example already discussed.

The coefficients w_i in equation (8.94) should be determined in such a way that for instance the following target variable is assigned to the two classes

$$z_j = \begin{cases} +1 \text{ for class 1} \\ -1 \text{ for class 2} \end{cases}$$

The individual matrix elements are given by:

$$\bar{c}_1 = \tfrac{1}{8}\,(-1 + 0.5 + 1.5 + 2 + 2.5 + 3.5 + 5) = \frac{7}{4} = 1.75$$

$$\bar{c}_1^2 = \tfrac{1}{8}\,(1 + 0.25 + 2.25 + 4 + 6.25 + 12.25 + 25) = \frac{51}{8} = 6.37$$

$$\bar{c}_2 = \tfrac{1}{8}\,(1 + 3 + 4.5 + 3 - 0.5 + 1 + 2 + 1) = \frac{15}{8} = 1.87$$

$$\bar{c}_2^2 = \tfrac{1}{8}\,(1 + 9 + 20.25 + 9 + 0.25 + 1 + 4 + 1) = \frac{45.5}{8} = 5.68$$

$$\bar{c}_1 \bar{c}_2 = \tfrac{1}{8}\,(-3 + 2.25 + 4.5 - 1 + 2.5 + 7 + 5) = \frac{17.25}{8} = 2.15$$

$$\overline{z_j c_1} = \tfrac{1}{8}\,(-1 + 0.5 + 1.5 - 2 - 2.5 - 3.5 - 5) = -\frac{12}{8} = -1.5$$

$$\overline{z_j \overline{c}_2} = \tfrac{1}{8} (1 + 3 + 4.5 + 3 + 0.5 - 1 - 2 - 1) = 1.0$$

$$\overline{z_j} = \tfrac{1}{2} (1 - 1) = 0$$

Thus

$$M = \begin{pmatrix} 1 & 1.75 & 1.87 \\ 1.75 & 6.37 & 2.15 \\ 1.87 & 2.15 & 5.68 \end{pmatrix} \quad \text{and} \quad s = \begin{pmatrix} 0 \\ -1.5 \\ 1 \end{pmatrix}$$

According to equation (8.97) this produces the following system of equations for the 3 required variables w_0, w_1 and w_2:

$$w_0 + 1.75\,w_1 + 1.87\,w_2 = 0$$

$$1.75\,w_0 + 6.37\,w_1 + 2.15\,w_2 = -1.5$$

$$1.87\,w_0 + 2.15\,w_1 + 5.68\,w_2 = 1 \ .$$

The solution is as follows: $w_0 = +0.11, w_1 = -0.36, w_2 = +0.28$.

Substituting in equation (8.94) gives the discrimination function $e(c) = 0.11 - 0.36c_1 + 0.28c_2$. This function is the best linear function of the features c_1 and c_2 which for all features of class 1 approximates as accurately as possible the value of 1 and for class 2 the value of -1.

If the feature vectors are substituted in the obtained discrimination function then for class 1: $e_1 = 0.39$: $e_2 = 1.31$; $e_3 = 1.19$; $e_4 = 0.41$ and for class 2: $e_1 = -0.75$; $e_2 = -0.51$; $e_3 = -0.59$; $e_4 = -1.41$.

The discrimination boundary lies at $e(c) = 0$, i.e. the equation of the dividing line is

$$0.11 - 0.36\,c_1 + 0.28\,c_2 = 0$$

or

$$-2\,c_1 + 1.5\,c_2 = -0.6$$

If one compares this result with the iteration method then it can be seen that the dividing line runs parallel to the line of the linear classifier shown in Fig. 8.36 (displacement of $+0.3$ on the c_1 axis).

Whereas the iterative method terminates when the two classes are separated no matter how even if the individual feature vectors are only just short of the dividing line, in the linear regression method the dividing line is positioned optimally in the sense of a root mean square between the two classes.

8.3.1.2 Implementation of a linear classifier

The discrimination rule expressed in equation (8.39) can be simply implemented in the software of a microcomputer (microprocessor).

Since often in industry no rigorous accuracy requirements are made of classifiers an analog implementation also suggests itself which usually reaches an accuracy of 0.5%.

The structure of a classsifier for the two-class problem is shown in Fig. 8.37 (8.2). The limiter produces the following output values:

$$A = +1 \text{ for } e(c) > 0 \text{ (class 1)}$$
$$A = -1 \text{ for } e(c) < 0 \text{ (class 2)}$$

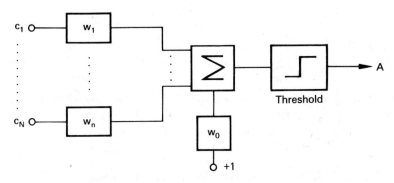

Fig. 8.37   Basic structure of the linear classifier (8.2).

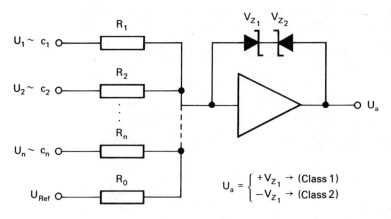

Fig. 8.38   Analog circuit for a linear classifier (8.38).

A circuit implementation is shown in Fig. 8.38. The feature vector to be classified is predefined as a voltage in terms of its components. The currents through the individual resistance are c_i/R_i. The resistances are chosen in such a way that they are inversely proportional to the components of the weighting vector: $R_i \sim 1/w_i$.

By adding the different currents an output voltage is produced which is proportional to the discrimination function equation (8.90).

The discrimination is made by means of an operational amplifier with 2 counter-connected Zener diodes providing feedback on the polarity of the output voltage (Fig. 8.38).

The classifier is cheap, reliable and fast. It needs about 1 ms for a discrimination.

8.3.2   Nearest neighbour classifier

It is intuitively obvious to make use of the spatial distance from one pattern to other patterns for classification. For instance in Fig. 8.39 it makes sense to assign the

103

pattern with the feature vector c to the class k_1 because of the spatial proximity.

We shall now derive the discrimination rule for a classifier which performs pattern classification based on the shortest spatial distance to a class. This classifier is called the nearest neighbour classifier.

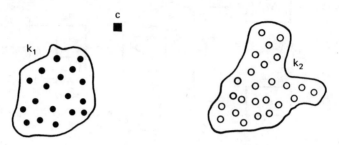

Fig. 8.39   Pattern classification based on spatial separation.

Fig. 8.40 shows the feature vectors of two classes and their geometrical separation for the two-dimensional case. The superscript in brackets denotes the class.

Vector algebra and Fig. 8.40 give the relationship:

$$c^{(1)} = c^{(2)} + e \qquad (8.100)$$

or

$$e = c^{(1)} - c^{(2)} \qquad (8.101)$$

The Euclidean distance between the two vectors is given by

$$|e| = \sqrt{\left[c_1^{(1)} - c_1^{(2)}\right]^2 + \left[c_2^{(1)} - c_2^{(2)}\right]^2} \qquad (8.102)$$

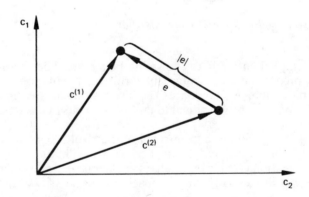

Fig. 8.40   Derivation of the nearest neighbour discrimination rule.

In the n-dimensional case the distance between the two vectors $c^{(1)} = (c_1^{(1)}, c_2^{(1)}, \ldots c_n^{(1)})^T$ and $c^{(2)} = (c_1^{(2)}, c_2^{(2)}, \ldots c_n^{(2)})^T$ where the equivalent row form of the vector is

104

used instead of the column form used previously indicated by a superscript T, is
given by:

$$|e| = \sqrt{\sum_{i=1}^{n} \left[c_i^{(1)} - c_i^{(2)} \right]^2} \; . \tag{8.103}$$

For a pattern to be classified the distance is obtained to every feature vector of
every class. A class is assigned to the pattern to which the nearest feature vector also
belongs. This is illustrated in Fig. 8.41.

The class k_1 contains six and the class k_2 contains four known feature vectors.
The pattern to be classified with the feature vector c has the smallest distance ($e_1 <
e_2$) to one of the feature vectors belonging to the class k_1. The pattern is therefore
assigned to the class k_1.

For $j = 1,2 \ldots ,m$ classes with $N_j = N_1, N_2 \ldots, N_m$ predefined (learnt)
n-dimensional feature vectors in each class there are a total of $N = N_1 + N_2 + \ldots N_m$
distances for a pattern to be classified with the feature vector c′:

$$|e|_j = \sqrt{\sum_{i=1}^{n} \left[c_i' - c_{il}^{(j)} \right]^2} \quad j=1,\ldots,m \tag{8.104}$$

$$1=1,\ldots,N_j$$

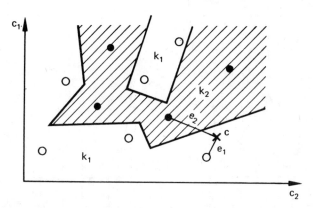

Fig. 8.41   Class assignment with the nearest neighbour classifier.

The discrimination rule is then:

$$|e|_j = \begin{cases} \text{minimum: pattern in class } j \\ \neq \text{minimum: otherwise} \end{cases} \tag{8.105}$$

8.3.2.1   Implementation of a nearest neighbour classifier
A computer (microprocessor) is most suitable for calculating the distances and for
making the classification using the discrimination rule (8.105). However for some
applications it may be advantageous to carry out the classification with an analog
circuit of the appropriate type.

As far as classification is concerned it is immaterial whether the distance is
calculated or its square. Then the square of the distance of the vector c′ to be

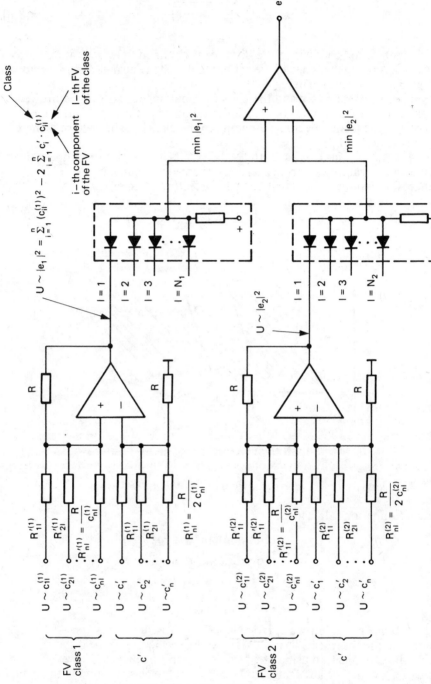

Fig. 8.42   Analog circuit for a nearest neighbour classifier.

classified to all other feature vectors can be obtained from equation (8.104) with some simplification of the form:

$$|e|^2 = \Sigma c'^2 - 2\,\Sigma c'c + \Sigma c^2 \tag{8.106}$$

If the square of the distance is calculated separately for the feature vectors in the two classes then the discrimination rule can be modified as follows:

$$e' = \min |\,e_1\,|^2 - \min |\,e_2\,|^2 = \begin{cases} <0, \text{if pattern in class 1} \\ >0, \text{if pattern in class 2} \end{cases} \tag{8.107}$$

with

$$|e_1|^2 = \Sigma(c^{(1)})^2 - 2\,\Sigma c'c^{(1)} \tag{8.108}$$

$$|e_2|^2 = \Sigma(c^{(2)})^2 - 2\,\Sigma c'c^{(2)} \tag{8.109}$$

The term $\Sigma c'^2$ is the same in both classes and so cancels out. This means that there is no quadratic term in the discrimination rule with respect to the pattern vector c'. A linear classifier is therefore involved.

Fig. 8.42 shows a possible analog circuit for the type of classification problem. For each of the N_1 known feature vectors of class 1 and each of the N_2 feature vectors of class 2 an operational amplifier is necessary with a resistance network consisting of 2n resistances, corresponding to the n components of the feature vector (FV). The output voltage of the operational amplifier is proportional to $|e_1|^2$ or $|e_2|^2$. A diode network with N_1 and N_2 diodes is suitable for determining the minima.

8.3.3 Nonlinear classifier

All the previous types of classifiers fail when it comes to problems where the features of one class are completely surrounded by those of another class. An example of such a case is shown in Fig. 8.43.

Here for instance a classifier suggests itself whose discrimination surface is shown in Fig. 8.44. In the two-dimensional case circles of radius r are drawn around given centres A (8.38). In multi-dimensional space the discrimination surfaces are hyperspheres.

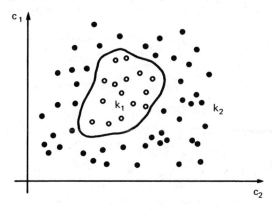

Fig. 8.43 Special arrangement of two pattern classes.

The square of the Euclidean distance of an n-dimensional vector being investigated to the centre point A with components A_i is given according to equation (8.104) by:

$$|e|^2 = \sum_{i=1}^{n} (c_i - A_i)^2 . \tag{8.110}$$

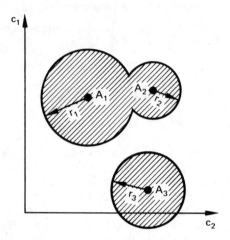

Fig. 8.44   Discrimination with a nonlinear classifier (8.38).

For $j = 1, 2, \ldots, m$ there are also m centres A_m.

With the definition

$$e'_j = r_j^2 - |e|_j^2 \tag{8.111}$$

the following discrimination rule is obtained

$$e'_j = \begin{cases} > 0 \text{ pattern in class } j \\ < 0 \text{ otherwise} \end{cases} \tag{8.112}$$

If $e'_j > 0$ the feature vector c being investigated lies inside the area around the centre of A. If $e_j < 0$ then the feature vector lies outside the predefined surface area around the centre A_j.

By multiplying out equation (8.110) one obtains with equation (8.111) with simplification of the form:

$$e' = r^2 - (\Sigma c^2 - 2 \Sigma Ac + \Sigma A^2) \tag{8.113}$$

The feature vector c is quadratic so that the classifier is nonlinear.

8.3.4   Sequential classifier
The classifiers described previously have the property that the number n of the components of the feature vectors is a fixed predefined value. The computing effort can be considerable if for instance with the nearest neighbour rule it is necessary to

calculate the distance of a high-dimensional feature vector to all stored feature vectors. Even when it has already been ascertained with great certainty that a feature vector belongs to a class on the basis of $n' < n$ components an unnecessarily large number of components still have to be included in the classification.

It makes more sense to develop a discrimination method which processes the components c_i of the feature vector step by step and a new component only involved if the certainty of recognition lies below a predefined threshold. This sequential classification reduces the computing effort sometimes considerably.

The basic idea of this method is to order the features according to their discriminating effect and to compare the features of the vector to be classified in this order (8.1).

The following discrimination rule can be formulated for the two-class problem (8.37).

The discrimination functions for the classes 1 and 2 are e_1 (c) and e_2 (c). The quotient of the discrimination functions is given by

$$S = \frac{e_1 \ (c)}{e_2 \ (c)} \tag{8.114}$$

Beginning with the first component c_1 the $i = 1,2 , \ldots , n$ components c_i of the feature vector being classified c are processed using the following discrimination rule:

$$c \text{ in class 1 if } S \geqslant S_1$$
$$c \text{ in class 2 if } S \geqslant S_2 \tag{8.115}$$
$$i = i + 1 \text{ if } S_2 < S < S_1$$

S_1 and S_2 are suitably chosen boundary values (so-called limit stops). Sometimes instead of the quotient the difference

$$S = e_1(c) - e_2 (c) \tag{8.116}$$

is used (8.39).

It is important for sequential feature processing to use the best feature components for discrimination at every step. In practice a histogram analysis is carried out for the selected features from which the discriminating power of the features can be ascertained (8.40, 8.41). The frequency of a certain value of the feature is recorded on the same diagram for each class. Fig. 8.45 shows an example of the frequency of occurrences of the values of a certain feature for different object classes. In this example all cases can be discriminated on the basis of this single feature. If this is not the case the frequency curves run into each other (Fig. 8.46). In general the discriminating power of a feature increases with increasing distance a between two classes (interclass separation) and with increasing distribution density within a class (intraclass separation). The classification of the object is carried out by a series of sequential discriminations the order of which is determined by the discrimination power of the features. The discrimination begins with features which because of their high discrimination power can lead to the classification result after a few steps in many cases.

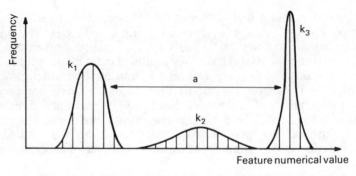

Fig. 8.45 Example of a histogram of a feature for different object classes.

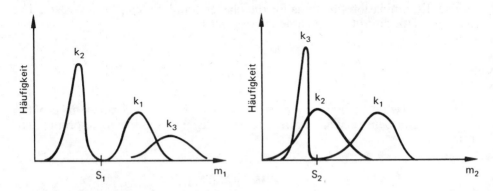

Fig. 8.46 Example of histogram analysis.

A simple example should illustrate the procedure. Fig. 8.46 shows the histogram of two features m_1 and m_2 for three classes. During classification the measurement results of the individual features are compared with thresholds which are obtained

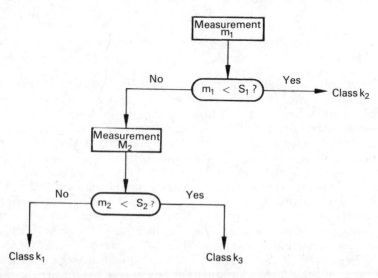

Fig. 8.47 Example of a decision tree for sequential classification.

110

from the histogram. The result of this discrimination determines how the classification continues. In this procedure the result is obtained by means of a so-called decision tree. The decision tree shown in Fig. 8.47 applies to this example.

Another example of sequential classification will be considered in Sections 10.1.7 and 10.1.11.

9. Coordinate matching between the optical sensor and the manipulation system

One of the most important tasks of an optical sensor is to determine target positions. Examples are the specification of the position of workpieces, markings or tracking curves.

One the target position has been determined it must be communicated to a manipulation system which is responsible for manipulating the workpieces. It should be pointed out here that in general the sensor and manipulation system use different coordinate systems. Preferably the sensor is located at the object and determines the required object coordinates with respect to its own coordinate system. The manipulation system also performs the movements in its own coordinate system (cylindrical, spherical, joint or Cartesian coordinates). This means that a coordinate transformation is necessary. In the simplest case this involves the addition of constants in the x-y plane. It is more complicated if randomly oriented objects have to be gripped with the help of a programmable manipulation system (industrial robot). Fig. 9.1 shows a diagram of such a manipulation system with examples of the possible movements (9.1). Fig. 9.2 shows the coupling of the sensor to the control system of a freely programmable manipulation system (9.2). In the absence of sensor control of the movements they are performed according to a fixed program. The position of the robot axes is monitored by the path measuring system. The sensor system controls the movement by the manipulation system on the basis of the measurement data provided by the sensor.

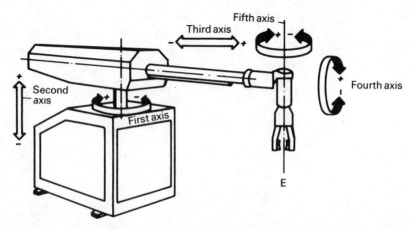

Fig. 9.1 *Example of programmable manipulation system (industrial robot) and the movement axes (9.1).*

112

The following considerations illustrate the kind of operations which have to be carried out for a transformation of the data measured by the sensor to the coordinate system of the manipulation system.

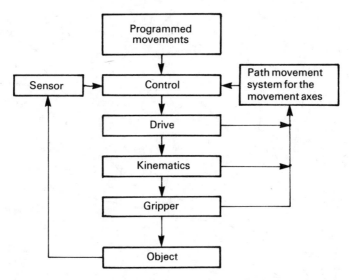

Fig. 9.2. Coupling of a sensor to the control system of a manipulation system (9.2).

The gripping of an object lying on a conveyor belt with the manipulation system shown in Fig. 9.1 will be given as an example (9.3). It will be assumed for the purposes of simplification that the workpiece can be gripped vertically from above. It is therefore not necessary to move the fourth axis which swivels the gripper; this axis can remain in the basic position during gripping. Axes 1, 3 and 5 are controlled from the sensor during gripping. The position of axis 2 which adjusts the height of the gripper is independent of the position of the object and is controlled by the manipulation system's drive control.

The task consists of determining the position of the end point E (Fig. 9.1) of the robot arm above the gripping point of the workpiece as a function of the data for this point measured by the sensor. In this example the end point of the robot arm is described in cylindrical coordinates: the direction of the arm is determined by axis 1 and the end position of the point E by the extension length of axis 3.

Fig. 9.3 shows the Cartesian sensor coordinate system and the cylindrical coordinate system of the industrial robot. For the purposes of simplification both systems should lie in the same plane in which they are however displaced and rotated with respect to each other.

The connecting vector between the two origins 0_1 (robot) and 0_2 sensor is defined by the distance c and the angle φ to the basic setting of the first axis.

The values of c, φ and the angle θ are measured at the system. For this the robot arm is driven to the origin of the sensor system of coordinates 0_2. The values for the distance c and the angle φ can then be determined directly using the path measuring system of the industrial robot.

Because of the rotational invariance the location of a workpiece is preferably characterised by the centre of gravity of the surface F (section 8.2.2.1). However in extremely rare cases a workpiece can also be gripped by the manipulation system at this point. The connection of the centre of gravity F to the possible gripping point is

113

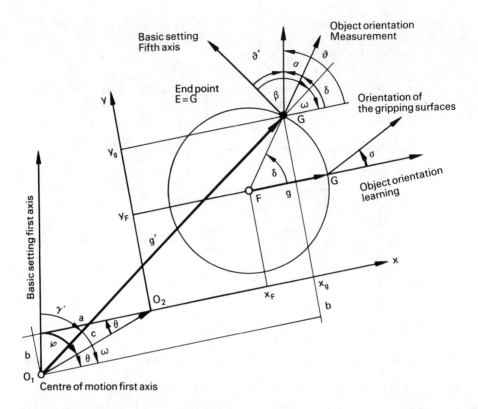

Fig. 9.3 Transformation of the Cartesian sensor coordinate system to the cylindrical coordinate system of the industrial robot (example) (9.3).

described in Fig. 9.3 by the vector g. In the case of a magnetic gripper the gripping point is defined with respect to the manipulation system at the centre point of the contacting magnet suface and in the case of a tong gripper as the centre point between the two open gripping tongs.

The gripping point G is not rotationally invariant and must be determined by the sensor with respect to a reference orientation. The reference orientation is the workpiece position input during learning. It is preferable here to align the workpieces with edges parallel to the axes of the sensor system. It is assumed in the following considerations for the purposes of simplicity that the gripping point vector g lies in the direction of the x-axis of the sensor system during learning (Fig. 9.3). If this is not the case then a constant angle has to be added on with regard to the gripping point G.

With this assumption the coordinates x_g and y_g of the gripping centre point G are given by

$$x_g = x_F + g_j \quad . \quad \cos \delta \qquad (9.1)$$

$$x_g = x_F + g_j \quad . \quad \sin \delta \qquad (9.2)$$

Here δ is the rotation of the object to be determined by the sensor with respect to the learnt reference orientation. The magnitude g_j of the connecting vector from the centre of gravity to the gripping centre point is a known constant for each orientation class j of the workpiece. The coordinates of the gripping point G can be determined with the centre of gravity coordinates determined by the sensor, the orientation class and the rotational angle.

When the gripping point has been found the gripper is rotated through an angle σ_j which is known and specific to the workpiece orientation class in order to be able to grip the workpiece in the required manner.

According to this the gripping surfaces have an angular orientation at the gripping point of

$$\vartheta = \sigma_j + \delta \; . \tag{9.3}$$

The transformation of the gripping point $G = (x_g, y_g)$ and the angle ϑ to the coordinate system of the manipulation system follows directly from Fig. 9.3. It is required to find the angle γ' which specifies the rotation of the first axis (Fig. 9.1) with respect to its basic setting as well as the extension length g' of the third axis and the gripper rotation ϑ' of the fifth axis with respect to its basic setting.

The rotation of the first axis (Fig. 9.3) is given by

$$\gamma' = \theta + \varphi - \omega \tag{9.4}$$

with

$$\omega = \arctan \frac{b + v \cdot y_g}{a + v \cdot x_g} \tag{9.5}$$

and

$$a = c \cdot \cos\theta \tag{9.6}$$
$$b = c \cdot \sin\theta \tag{9.7}$$

Here v is the ratio of the numerical values which represent a predefined length unit in the manipulation system and the sensor system (scale matching).

The required extension length g' of the third axis can be obtained from Fig. 9.3 (9.3):

$$g' = \sqrt{(a + v \cdot x_g)^2 + (b + v \cdot y_g)^2} \tag{9.8}$$

The gripper rotation of the fifth axis is given by (Fig. 9.3).

$$\vartheta' = \beta + \omega - \vartheta \tag{9.9}$$

Here β is the angle between the arm of the manipulation system pointing to the gripping point and the basic setting of the fifth axis. This angle is constant and known.

The required values γ', g' and ϑ' have to be finally multiplied with a scaling factor which relates these values to the corresponding movement increments of manipulation system. Furthermore there are constants to be subtracted because the arm of the manipulation system already has a certain extension in the starting position.

115

10. Examples of sensor implementations in industry

The foregoing chapters have considered the individual components of optical, acoustic and tactile sensor systems as well as the principles of signal processing.

Using this as a basis we shall now present sensor systems which have been implemented in practice. The literature on the implementation of intelligent measuring systems is scattered throughout the various meeting reports and specialist journals. Very often these publications cannot be understood by those not working in pattern recognition because of lack of specialised knowledge so that it is necessary to familiarise oneself with the primary literature. The knowledge necessary for the examples presented here is contained in the previous chapters so that the book as a whole is self-contained.

The examples serve principally to illustrate the applicability of the basic principles considered earlier. They are intended to stimulate the engineer entrusted with automation tasks and bring his attention to possible solutions. At the same time these examples show the difficulties and limitations of what is feasible today.

Corresponding to the frequency of the tasks much emphasis is given to the

Fig. 10.1   Workpieces for ordering tasks.

116

production sector. Especially in small- and medium-scale manufacture there is the desire to humanise and rationalise the working operations (10.1, 10.2, 10.3). Industrial robots are suitable for many manipulation tasks (Fig. 9.1). These manipulation systems are freely programmable in many axes and perform a certain operation with grippers or tools (10.4). Some of the following examples are sensor systems for controlling such industrial robots.

The classification of sensor tasks in Chapter 3 has shown that the main sensor tasks are of the pattern recognition type (10.5).

The selected examples for optical and tactile sensors can be mainly assigned to the following working areas:

(a) Ordering tasks occur during handling workpieces which are transported unordered in bunkers inside the factory but which must be ordered and in a known position in order to be gripped by manipulation systems. With small batches purely mechanical ordering equipment is excluded on economic grounds. It is too expensive to adapt the same mechanical ordering equipment to different types of workpiece (Fig. 10.1) so that the various geometrical features can be taken into account, e.g. with the help of mechanical baffles.

(b) Target finding in assembly tasks where the position of holes, threads etc. on components on assembly belts or at assembly sites are to be determined in order to automatically insert bolts, sockets, pins, screws, etc.

(c) Target finding in order to position tools by determining the target itself or by determining reference points which are already present (holes) or which are specially marked.

(d) Position measurement for gripping workpieces which are roughly positioned on universal pallets, suspension gear of chain conveyors or on conveyor belts, e.g. in the plating and painting industry.

(e) Visual inspection tasks in quality control.

The acoustic sensor examples come from quality control and the surveillance of installations.

The pre-eminence of optical systems is reflected in the large number of examples.

10.1 Optical sensor systems
10.1.1 Recognition and position determination of objects by optical correlation
The principle of an incoherent optical correlator which is used to compare the object concerned with a stored image is described in Section 8.2.2.2. We shall now present the implementation of such a correlator for recognition and position determination linked to an industrial robot (10.6). The key importance of this development is that it is the first German sensor system for controlling an industrial robot.

This correlator is restricted to the recognition and measurement of circular patterns which makes it very simple. Investigations on workpieces have shown that about 66% of the tasks involve rotating parts (10.7). For instance in assembly cylindrical parts such as screws, bolts, pins, sockets, rings, axles, etc. are inserted into circular target positions.

The basic structure of the correlator is shown in Fig. 10.2. The purpose of the equipment is to determine the position of a round pattern in the x-y plane.

The pattern uniformly illuminated from a light source is focussed on to a rotating disc by the lens. Comparison masks in transparent form are located on the circumference of the rotating disc. The light flux passing through the masks is focussed on to a photoreceptor by the lens. When the pattern moves in the x-direction the orbit of the masks on the rotating disc. Depending on the geometrical conditions the individual masks are overlapped (correlated) by the pattern image. The x position of the pattern is determined by the point in time of maximum overlap.

In order to determine the y position of the pattern the angular orientation of the mask is measured at optimum overlap. This is done with the help of a shaft encoder linked to the rotating disc. The y-coordinate is then calculated from:

$$y = \frac{g}{b} \cdot R.\sin \varphi$$

(10.1)

where g/b is the lateral amplification (object to image width) and R is the distance of the masks from the centre of rotation.

As a rule the masks are applied photographically to film material. There are two possibilities for the form of the comparison masks. One is to have black circles on a light background and the other is the complementary case with a transparent circle on a non-transparent black background. In the first case the signal of the photo-detector is a maximum when the pattern and mask overlap and in the second case it is a minimum.

This set of masks has to be extended if interfering outside patterns come into the picture field in addition to the required round patterns. In practice this is nearly always in the form of fractures, slits, beginning and end of components, shadows of edges or holes with different diameters, etc. The task is then extended from pure position measurement to pattern recognition.

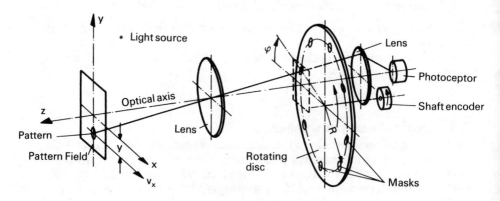

Fig. 10.2   Principle of the optical correlator (10.6).

The pattern range can be divided into two classes:

(1) Circles with the required nominal diameter D_o (Class 1).
(2) All other patterns including circles with $D \gtrless D_o$ (class 2).

If two circular ring masks are chosen as shown in Fig. 10.3 then the classes can be distinguished in a simple manner. The inner and outer masks are placed in pairs behind each other on the rotating disc. This does not essentially change anything in the method of position determination. The inner mask replaces the previous full circle mask for the measurement. The class discrimination can be made on the basis

118

of the correlation signals from the inner and outer circular masks: the detection of a specific signal configuration indicates the presence of the required pattern.

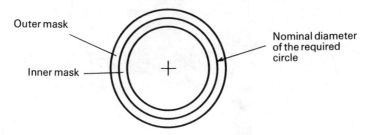

Outer mask

Nominal diameter of the required circle

Inner mask

Fig. 10.3   Pair of masks for recognising a predefined circle diameter.

The accuracy of position measurement of the y coordinate for moving sheet metal parts (conveyor belt speed 60 mm/s) is ± 0.4 mm for a hole diameter to be recognised of 8.7 ± 0.2 mm (10.6). This accuracy is adequate for a link-up with an automatic manipulation system. Usually the accuracy of such industrial robots is of the order of 1 mm.

Fig. 10.4 shows a diagram of the link-up of an optical correlator with an industrial robot. Parts of all shapes run underneath the correlator on the conveyor belt. The correlator recognises the round parts with the required diameter, measures their position and communicates the measurement result as well as another signal on the presence of the workpiece to the control system of the manipulator. This grips the workpiece and brings it to a processing station.

Fig. 10.5 shows a photograph of the whole arrangement. The correlator with the illuminating equipment is at the top. The gripper which can be seen on the right only seizes parts which have been correctly recognised. These can arrive in any position on the conveyor belt.

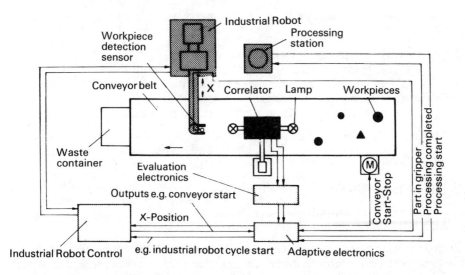

Fig. 10.4   Link-up of the optical correlator with a programmable manipulation system (10.6).

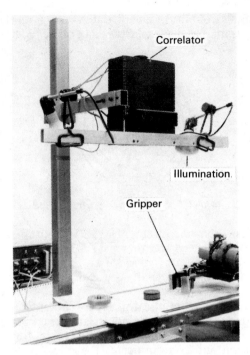

Correlator

Illumination

Gripper

Fig. 10.5   Optical correlator for controlling an automatic manipulation system (Source: IITB and IPA Stuttgart).

Although the sensor for the frequently occurring and important task of recognising and measuring the position of circular patterns has been developed especially for this task it can still be adapted for related tasks.

If the pattern diameter changes the rotating disc can be replaced with another one with new reference patterns. Frequently however it is sufficient to change the lens or the focussing conditions by altering the geometry.

10.1.2   Television camera for position determination

As discussed in Section 8.2.2.1 an area centre of gravity is suitable for characterising the position of an object. If an object is scanned with a television camera then the area of gravity is defined by equations (8.51) and (8.52).

There are in principle two possibilities for determining the centre of gravity:

1. Processing the information directly during the line scan (on-line processing).
2. Storing the contour points of the object and subsequent processing (off-line processing).

On-line processing during line scanning means that the result is readily available and is used when the processing time has to be very short. This principle is implemented in the following example of a television camera for position determination.

The block diagram is shown in Fig. 10.6 (10.8). The video signal from the television camera is led to a digitising circuit in which a binary video signal is generated by means of a threshold in the voltage level (Section 4.5.2). This controls the individual steps of the computer unit CU via the transfer logic. The data obtained there (centre of gravity coordinates x_F, y_F, area F, maximum extensions

x_{max}, y_{max} in accordance with Section 8.2.2.1) are supplied to the display unit via a bus system and to the robot control system and a comparator via a D/A converter. Boundary values for the recognition criteria F, x_{max}, y_{max} are preset in an input device and compared in the comparator. A character generator is used to display the obtained centre of gravity coordinates x_F, y_F at the monitor. The sequence control fixed in a PROM is provided with a learning and operating mode which allows the correct object to be learnt (teach-in phase).

During this learning phase the television camera is trained on the required pattern and the area as well as the maximum extensions x_{max} and y_{max} are determined. These three variables are adequate for the recognition of many work-pieces.

In the operating mode the measured values for these three variables are compared with the stored data. According to the result there is a yes/no display indicating whether the learnt pattern is present or not. At the same time the centre of gravity coordinates are passed to the robot control system.

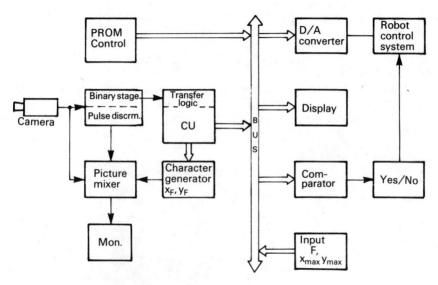

Fig. 10.6   Block diagram of the sensor system (10.8).

Fig. 10.7 shows the block diagram for the high-speed computer unit CU. The double sum in the numerator of equation (8.51) for determining the centre of gravity coordinate x_F is continuously obtained over the entire picture by squaring the values from the counter C_1 (which counts every line from beginning to end) in the multiplier MULT and forming a running total in the adder ACC1. The number of equation (8.52) for the centre of gravity coordinate y_F is determined by adding the section lengths m of each line in an electronic counter C_2. The line number n is kept in a counter C_3. At the end of every line while the electron beam of the camera jumps to the next beginning of a line the values of C_2 and C_3 are multiplied according to equation (8.52) and added to the running total in the adder ACC2. The denominator of equations (8.51) and (8.52) which corresponds to the area F of the pattern is formed in the counter C_4. In order to do this the counter is started at the dark-light transitions and stopped at the light-dark transitions. The section lengths

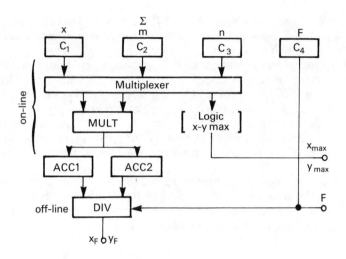

Fig. 10.7   *Block diagram of the high-speed computer unit (10.6).*

are determined from the number of pulses counted between these two boundaries (Section 4.5.2, Fig. 4.22).

Whereas the numerator and denominator of x_F and y_F are formed on-line, i.e. during the television scanning the subsequent division is performed off-line in the divider. At the same time the maximum object extensions x_{max} and y_{max} are determined in a simple circuit.

Fig. 10.8   *Laboratory assembly of the sensor system (Source: IITB).*

Fig. 10.8 shows the laboratory assembly of the sensor system with a pattern shown on the television monitor whose centre of gravity is marked by an electronically blended cross.

The measuring accuracy which can be obtained essentially depends on the following parameters:

– Quality of the television camera (geometrical faults, stability)
– Scanning of the binary picture
– Pattern contrast

On the average the measuring error for the position is about ± 2% of the picture area being processed.

The measurement is made every second raster (field), i.e. every 40 ms (cf. Section 4.1.2: picture frequency 25 Hz). The system is therefore suitable for measuring patterns which are not moving too fast.

The application of the sensor in determining the start of the weld for arc welding will now be described.

Welds and the start of welds represent patterns which are difficult to process optically (Section 4.5.1, Fig. 4.20). It is therefore desirable to measure well-defined and easily recognisable reference points in the vicinity of the start of weld and so determine its position.

The function of the sensor is therefore to measure reference points for the start of weld such as holes, apertures, paint marks etc. which are easier to recognise than the weld itself. Here one makes the simplifying assumption that the welds are only displaced by translation so that there are no rotational tolerances. Making this assumption the seams can be welded automatically following a program after the start of weld has been found. If the weld is displaced the reference mark is also displaced.

Fig. 10.9 shows a diagram such as a sensor-controlled welding robot. Fig. 10.10 shows the system in practical use. The gripper of an industrial robot guides the

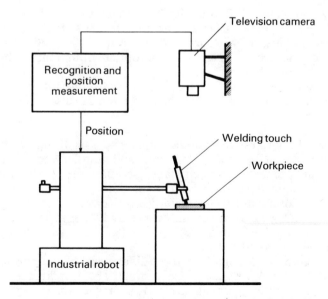

Fig. 10.9   Diagram of a sensor-controlled industrial robot for arc welding.

torch according to the measurement data taken by the television camera and processed by the sensor. The workpiece to be welded is illuminated diagonally by two projectors. The camera is protected by a swing shutter against the bright light of the welding. The binary picture cannot be distorted by the light of the welding process since the measurement is performed before the actual welding.

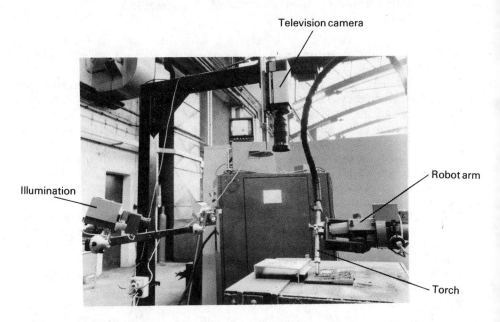

Fig. 10.10   Determination of the start of weld: sensor for the torch control (10.9) (Source: IWKA).

There are problems in illuminating the reference point. When the illumination is oblique (Fig. 10.10) the surfaces around holes reflect so that they then appear as a dark pattern against a light surround. Since the holes and apertures are often randomly distributed in space the appropriate illumination has often to be determined by trial and error.

Better results are obtained by making a fluorescent paint marking since this is unaffected by the direction of the light source and can also be better distinguished from interferences.

10.1.3   Recognition system for orienting parts with small feature differences

The application of automatic handling and transfer devices for automating assembly and processing often fails because the workpieces cannot be automatically delivered in the correct orientations.

Parts to be ordered very frequently only possess extremely small feature differences e.g. cogs, bezels, cones, recesses, reducers, threads. Parts of this type occur very often in the form of hinge pins, chain studs, setscrew blanks, shafts for small-size gearboxes and motors.

The recognition of small feature differences by humans requires high concentration and is physically exhausting. The importance of a sensor for distinguishing between parts can be clearly seen when the workpieces are handled in bulk and aligned along one axis arrive at a processing station by conveyor belt.

However fully automatic handling fails when it comes to recognising the orientation.

The following example of an optical sensor for solving this problem (10.10) shows that it is also possible to solve complicated problems in recognition with relatively simple means.

The measuring system consists of a light barrier with scoring logic. The part whose orientation is required is guided on to a rail (slide, conveyor belt, spiral of a vibrator conveyor) between the transmitter and receiver as shown in Fig. 10.11. When the workpiece interrupts the infrared beam the voltage at the receiver diode changes. The voltage rises in proportion to the surface covered and depends on the shape of the workpiece. A typical voltage curve is shown in Fig. 10.12. The section between the two levels is used for recognition. At the beginning of the workpiece transit a counter is started when level 1 is reached and stopped when level 2 is reached. The result of counting is stored. At the end of the transit on reaching level 2 the counter begins to count backwards until level 1 is reached. The final result in the counter depends on the orientation of the moving part.

In this example the feature vector only has one component which is determined by the result of counting.

Since the times t_1 and t_2 are only compared to one another the result is independent of the transporting speed insofar as this is constant during the transit of the part. If a new part has to be recognised it is only necessary to reset the two levels.

By using an infrared transmitting diode the sensor is made insensitive to outside light influences.

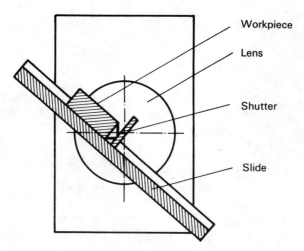

Fig. 10.11   Principle of the recognition device for small feature differences (10.10).

Fig. 10.13 shows the sensor system for recognition of orientation combined with a vibrator conveyor. Parts whose orientation has been recognised as correct continue being transported and those with incorrect orientation are returned to the delivery bin.

The smallest features to be recognised involve length dimensions of the order of 0.5 mm. The transit speed depends on the length of the workpiece. The shortest transit time for reliable recognition is 20 ms. For a 20 mm long workpiece this gives

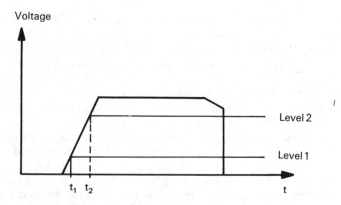

Fig. 10.12   Evaluation of the voltage at the receiver diode.

a maximum speed of 1 m/s. If there is a gap of 30 mm between individual work-pieces there is a throughput of 20 pieces/s.

This procedure works independently of the workpiece material. For example it can also be used with glass ampoules.

Fig. 10.13   Example of an ordering line for the delivery of bolts in correct orientations on a vibrator conveyor (10.10) (Source: BOSCH).

Another application area for the sensor is the integrity check. Faults are detected if the measured contour area deviates by about 5% from the required area. Fig. 10.14 gives a summary of the various inspection tasks which can be carried out.

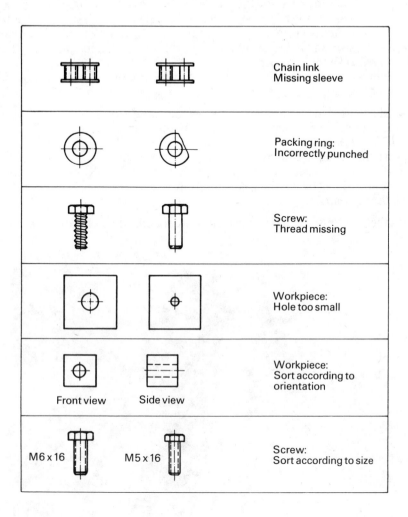

Fig. 10.14   Examples of inspection tasks (10.9).

10.1.4   Programmable workpiece recognition system

With the present state of development flexible automatic handling systems (industrial robots) are not capable of performing one of the most important tasks in handling, the orientation of parts. In order for industrial robots to be economically viable they must be provided with flexible orienting systems which bring the disordered workpieces automatically to a defined position and orientation. A simple, flexible orienting system for solving this problem will now be described (10.11).

Classifying this method as an optical, acoustic or tactile procedure as done previously presents some difficulties as the transducers used are inductive and therefore not optical. There are several reasons why this method has all the same been classified as an optical procedure. On the one hand, optical methods concern themselves exclusively with tasks which humans perform with the help of their visual faculties. The task of ordering (orienting) also falls under this heading. The term optical should not be understood exclusively in the physical sense here. It should more generally describe the whole of the task which humans can tackle due to their visual capabilities. On the other hand the signal processing involved in this method with inductive transducers is entirely identical to that of optical methods (cf. following sections). The principle of a programmable recognition system can be clearly seen from this example.

The following technical specifications for an ordering system can be derived from the objective outlined above.

- Suitable for as many different workpiece types as possible.
- No mechanical devices used which are difficult to refit.
- Retooling by easy reprogramming (software adaptability).
- Low hardware requirements.
- Insensitivity to industrial environmental factors.
- Simple and robust mechanical construction.
- High throughput.

With the present state of development the expenditure on hardware can only be

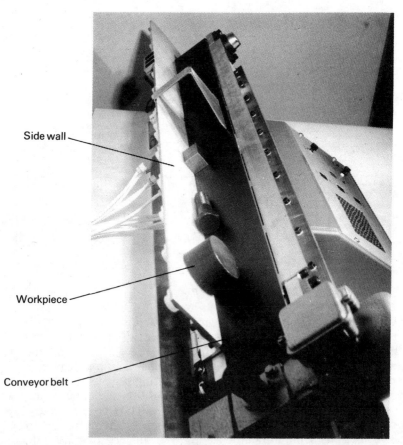

Side wall

Workpiece

Conveyor belt

Fig. 10.15 *Workpiece alignment in the recognition station (10.11). Source: IBP Pietzsch).*

kept within reasonable limits by reducing the number of degrees of freedom (Section 4.5.3).

In this example a V-shaped channel was chosen as a flexible mechanical solution. The workpieces can only adopt a few possible discrete orientations under the influence of gravity. The V-shaped transport channel is formed by the belt of a conveyor and a rigid side wall with good friction properties (Fig. 10.15). Inductive proximity switches are arranged behind the belt and side wall as shown in Fig. 10.16. They scan the workpiece simultaneously in two planes. The switching gap of the inductive transducers is larger than the thickness of the belt and side wall. When the workpiece moves, a grid picture is generated of the surfaces turned towards the switches from the signals, which at first only correspond to one picture line by means of repeated interrogation of the proximity switches at programmed intervals. The grid picture shown in Fig. 10.17b is obtained from the workpiece shown in Fig. 10.17a for example. The grid picture gives a coarse contour image of the workpiece. L means that a contour was found at the corresponding proximity switch.

The components of the feature vector are given by the binary result of the proximity switches.

For workpiece recognition the grid picture is compared with the reference pictures stored during programming in logic circuits. The workpiece with the largest amount of agreement is selected.

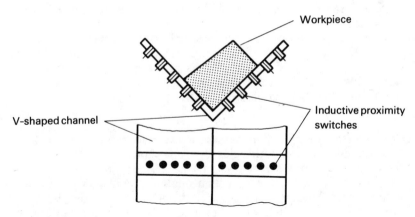

Fig. 10.16   Arrangement of the inductive proximity switches (10.11).

For the programming the workpiece is allowed to pass by the inductive switches in the V-channel in the required orientation and so assigned an orientation class at the operating console for each orientation. This orientation number is electronically stored together with the grid picture. The programming is a simple teach-in process and does not require any special knowledge. During the recognition process the signals emitted by the array of switches are electronically compared with those of the stored reference pictures. If the switch signals agree in all points with a reference picture or only deviate by a small number of points preset at the operating console then the workpiece is recognised. The orientation number is displayed at the operating console as the result of the recognition process and is available as an electrical signal e.g. for calling an appropriate robot program. Workpieces with images showing larger deviations from each of the reference

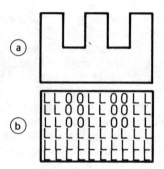

Fig. 10.17   Example of a grid picture (10.11).
(a) Workpiece
(b) Grid picture of the workpiece.

pictures are rejected as useless by a built-in deflector. This adjustable tolerance limit is necessary to allow for small deviations in the geometry of the workpieces.

Possible applications are the loading of processing machines with parts arriving in disordered fashion or the depositing of different parts on appropriate pallets. The recognition system can also be used in quality control for testing workpieces.

The system is limited to recognising metal workpieces through the use of inductive switches. However, non-metallic objects can also be recognised by using capacitive switches.

The advantage of this recognition system is its robustness and its insensitivity to harsh environmental conditions. Its disadvantage is a limited resolution of the grid image. Because of the geometrical dimensions of the proximity switches their spatial packing density is restricted so that only coarse scanning is possible. The following television system overcomes this disadvantage.

10.1.5   Television sensor for position measurement and for recognising workpieces with discrete orientation classes in one plane.

In the two previous sections sensor systems have been presented where the recognition of workpieces was achieved with a small outlay in terms of hardware. However, in most cases of industrial recognition the problem is so complicated that it can only be solved with the help of computers. Only systems are suitable here which can manage with inexpensive microprocessors and a relatively small memory. A further important requirement is a short computing time which must be considerably below one second.

A picture sensor will now be described which by making a considerable reduction in the picture information leads to a simple and inexpensive system for workpiece recognition (10.12, 10.13).

As described in Section 4.5.3. the isolated workpieces lying on a conveyor belt are brought against a mechanical stop in order to limit the number of degrees of freedom. This produces a limited number of discrete and stable workpiece orientation classes. The conveyor belt is transparent so that when light is shone through a binary contour image of the workpiece is obtained.

The contour image of the workpiece is scanned with a conventional television camera placed vertical to the plane of the belt with 625 lines per picture. However, only a few lines of the television picture are used for the actual picture recognition. In Fig. 10.18 for example there are lines z_1 and z_2. They generate the coordinate

130

values x_{ik} with respect to a starting mark where i is the serial number of the selected line and k the serial number of the binary changes in brightness occurring along this line. The values x_{ik} obtained in this way are used to form the differences from the starting value x_{11}:

$$x'_{ik} = x_{ik} - x_{11} \qquad (10.2)$$

In this way one in independent of the distance of the workpiece from the starting mark. The set of all differences gives a characteristic feature vector for each workpiece orientation:

$$c = (x'_{12,}\ x'_{13,\dots,}\ x'_{ik}) \qquad (10.3)$$

the components of which are the section lengths of the scanned black-white transitions.

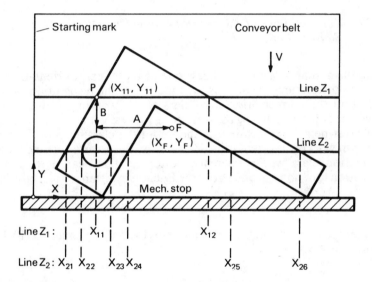

Fig. 10.18   Principle of workpiece scanning with a few selected television lines (10.12).

Fig. 10.19 shows a photograph as shown on the monitor of a workpiece at a stop (bottom edge). Starting from a reference mark (lefthand edge) the workpiece is scanned along four television lines. In order to show up the section lengths they are automatically scanned light or dark depending on the background.

It turns out that four lines are adequate for characterising most workpieces.

When the picture sensors are programmed the feature vectors c_R serving as references are obtained and stored in memory for all workpieces. For this the workpiece is presented to the television camera on the conveyor belt at the stop in every possible orientation. The workpiece contour, together with the lines marked by light scanning, are displayed on a monitor. The position of the individual lines can be shifted electronically and are appropriately selected by an operator. The line selection is the same for all orientations of a workpiece. It is possible to automate this process. The monitor is only needed during programming and is otherwise free for other tasks, for instance for programming other sensors.

131

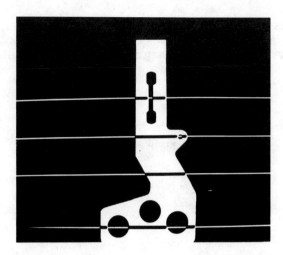

*Fig. 10.19   Monitor image of a workpiece scanned with four television lines (10.13)
(Source: IITB).*

During the actual measuring process one of these learnt orientation classes is presented to the picture sensor. It is the function of the recognition system of the sensor to find the correct orientation class of the workpiece by comparing the feature vector c obtained during the measuring process with all reference vectors c_R.

The nearest-neighbour classifier is chosen for the classification (Section 8.3.2). An unknown workpiece orientation described by the feature vector c is assigned to the orientation class whose reference vector c_R it is nearest to. For this the discrimination function e formed from the Euclidean distance must be minimal:

$$e = \sum_{ik} (c - c_R)^2 = min \qquad (10.4)$$

For easier technical implementation it helps to simplify equation (10.4) without any practical restrictions to:

$$e' = \sum_{ik} (c - c_R) = min \qquad (10.5)$$

The magnitudes of the differences of the vector components are summed and the orientation class selected for which the discrimination function has a minimum.

If the value of the discrimination function exceeds a predefined threshold the workpiece is rejected as unknown and presented to the recognition system again in another orientation.

If the orientation of the workpiece on the conveyor belt is known, then any fixed intersection point of one of the television lines with the workpiece will give its position at the same time. The x-coordinate of the intersection point is measured from the distance to the starting mark. The y-coordinate is then specified from the known workpiece orientation or can be determined from the line number and the known line number and the known line spacing.

If for instance the first point of intersection of the line Z_1 is chosen as the reference point (point $P = (x_{11}, y_{11})$ in Fig. 10.18) then each point of the workpiece can be calculated by addition of two orientation-specific constants x^L, y^L. For instance the orientation-specific constants with respect to the centre of gravity $F = (x_F, y_F)$ are given by the section lengths $x^L = A$ and $y^L = B$. According to this the coordinates of the centre of gravity are $F = (x_F, y_F) = (x_{11} + A, y_{11} - B)$. The gripper and gripper arm of the industrial robot can now be controlled with the orientation class and position known.

The two operations are characteristic for this procedure: the measurement of the length section x_{ik} and the determination of the orientation class with the help of the classification algorithm. This twofold division is reflected in the hardware implementation. The measurement data on the workpiece contours are determined by a high-speed counting circuit during one line scan and subsequently processed by a microprocessor.

Fig. 10.20 shows the block diagram of the sensor. The PBS video signal is converted to a binary image and after synchronising separation controls the counter via a logic circuit. Here the workpiece contours are processed during line scanning according to the television lines selected by the operator and read into a high-speed RAM. Because of the data reduction a capacity of the measurement memory of 64×8 bits is adequate, i.e. a feature vector can have up to 64 components with an accuracy of 8 bits. The measurement sequence (RAM clearance, memory housekeeping) is coordinated by a measurement sequence control system, the basic modules of which are a PROM program memory, status counter, conditional multiplexer and function decoder. On completion of the measuring process the data are transferred from the RAM to the microprocessor system and processed there according to the classification algorithm.

With m orientation classes to be distinguished the memory requirement for the measured feature vector and the m reference vectors is $(m + 1) \times 64 \times 8$ bits which with 15 orientation classes means a memory requirement of 1 k Byte. Another 0.5 k Byte is needed for the classification algorithm program. The reduction in the amount of information is especially clear here. Whereas the information content of a single binary television picture is about 5.10^5 bits only 8.10^3 bits of memory capacity are needed for processing 15 pictures of the corresponding orientation classes using this procedure. The total time for picture processing is about 70 ms.

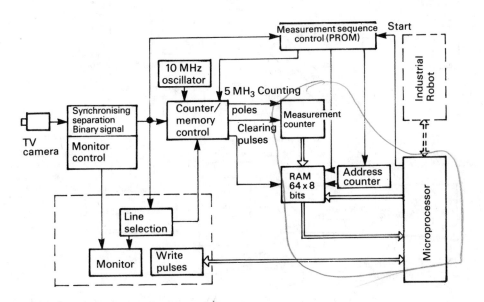

Fig. 10.20   Block diagram of the sensor system.

133

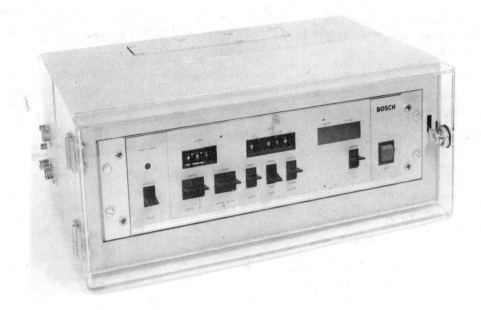

Fig. 10.21   View of the sensor system (10.14) (Source BOSCH).

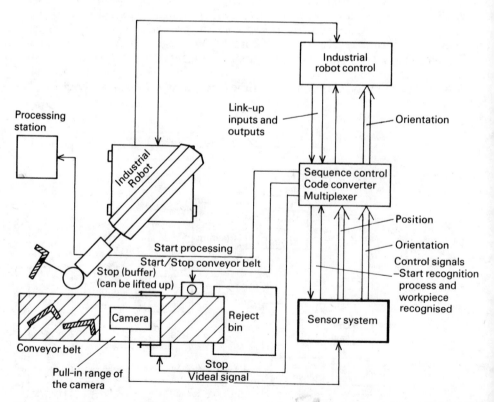

Fig. 10.22   Link-up of the television sensor with an industrial robot for ordering
workpieces (10.14).

Fig. 10.21 shows the overall view of the sensor with display and operating elements. The front panel contains all the operating controls necessary for programming as well as a display for the orientation class and position. The total system including the microprocessor only needs a single 19″ rack. This example makes it clear that sensor systems in industry do not have to be elaborate.

The link-up of the sensor system with an industrial robot for fully automatic picking and placing workpieces in a processing station will now be described. The principle of the overall system is shown in Fig. 10.21.

The workpieces to be ordered (shown in Fig. 10.22 as brackets) are brought by a conveyor belt against a stop arranged in the picture field of the camera. If the television sensor recognises the workpiece orientation the conveyor belt is stopped and the orientation class and position of the workpiece communicated to the control system of the industrial robot. The robot grasps the workpiece and restarts the conveyor belt after the gripper has left the picture field. The workpiece is deposited at a work station (e.g. a press). In the meantime a new workpiece has arrived at the top and the process is repeated. If a workpiece is not recognised the stop is lifted up and the workpiece transported to a reject bin.

Fig. 10.23 shows the entire arrangement in practice. A transparent conveyor belt is illuminated from the rear and a binary contour picture of the workpiece thus generated. The considerable soiling of the conveyor belt seen in Fig. 10.23 as always, occurs in practice, can be dealt with even in the most unfavourable cases by a suitable threshold setting. Fig. 10.24 shows the monitor picture of a workpiece scanned in the light passing through a soiled conveyor belt with four lines. Only the workpiece contours are processed for the evaluation, as can be seen by the light-scanned lines at these points. Even with considerable soiling the background does not introduce any faults.

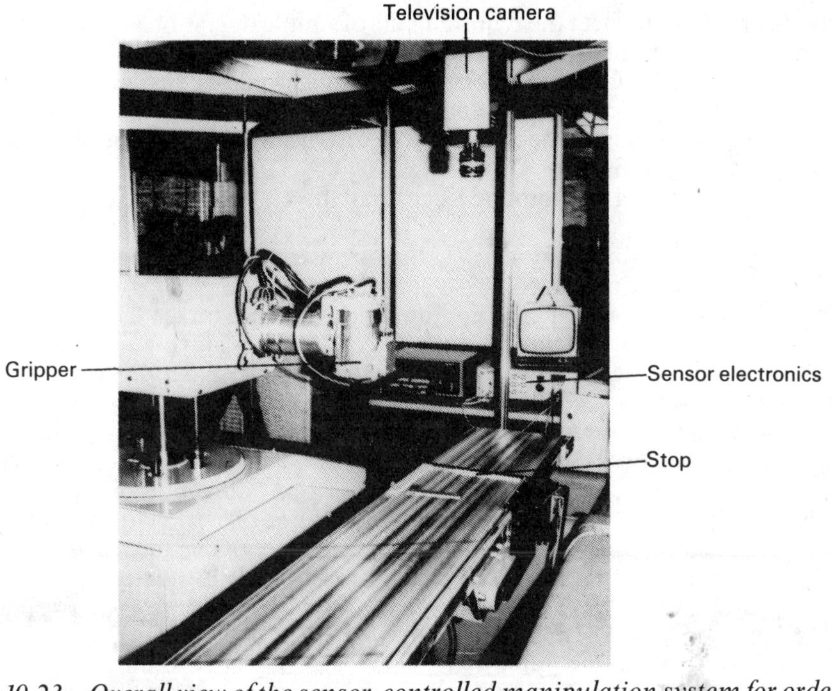

Fig. 10.23   Overall view of the sensor-controlled manipulation system for ordering workpieces (10.14) (Source: IPA Stuttgart).

135

Fig. 10.24   Generation of a contour picture even with soiled background (Source: IITB).

10.1.6   Television sensor for position measurement and recognition of workpieces with rotational degrees of freedom in one plane

In the previous example the workpieces are aligned with mechanical aids in such a way that only a few discrete orientations are possible.

A sensor will now be presented which determines the position and orientation mode of a workpiece with any orientation in one plane (conveyor belt) (10.15, 10.16).

It is only possible to meet the requirements of simple operation, short processing time and low cost if limitations are taken into account:

(a) The objects are deposited on one plane which is perpendicular to the optical axis of the imaging device and has a defined distance to it.

(b) Single presentation of the objects by means of a mechanical appliance which is not workpiece-specific.

(c) Illumination and background such that the contour of the object is fully imaged (back-lighting).

(d) Alternatively illumination such that in direct light characteristic binary workpiece pictures are produced.

(e) Flash illumination which in conjunction with the storage facility of a television camera avoids any distortion of the object while it is moving.

(f) Avoidance of perspective distortions with three-dimensional objects by making the distance between imaging device and object sufficiently large.

If several objects are present in the picture field it is assumed that the objects are isolated and behind one another on the conveyor belt. The lines of the television camera for imaging are perpendicular to the transport direction of the conveyor belt. This arrangement ensures that there is at least one television line between two objects. These empty lines are detected and in this way the picture field is divided up. The picture field zones thus obtained are successively evaluated starting from the top edge of the picture. Zones with objects which touch the edge of the picture or whose area falls below a certain adjustable value are not considered.

For an object selected according to this preprocessing the type and rotation of the object is computed with respect to an object point. The surface centre of gravity of the binary picture serves as the object point since its position is rotationally and

136

10.25 Photograph of the workpiece (10.15).

translationally invariant with respect to the contour (Section 8.2.2.1). The area is calculated from equation (8.48) and the centre of gravity from equations (8.49) and (8.50).

In the interests of short processing times the numerators and denominators of these equations are determined in a similar manner to the sensor described in Section 10.1.2 with the help of digital circuits.

The type, orientation and rotation of the object are obtained by means of a feature comparison. In a learning phase translationally and rotationally invariant shape features and orientation features specifying the position of retation are stored in memory from the binary picture. Shape features are area, number of corners, number of holes, etc. Integral features such as the area, moments and perimeter make recognition more likely in the event of distorted pictures than differential

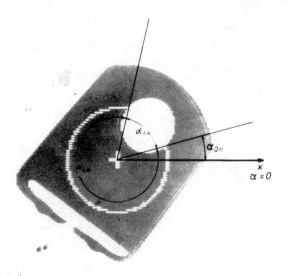

Fig. 10.26 Measuring variables for the polar check (10.15).

137

features such as corners, maximum radius, and holes since distortions partly cancel each other out or only affect the integral value to a small extent. By the same token small shape features such as bores, teeth, or indentations only produce small changes and are, therefore, more difficult to recognise when processing integral features.

The recognition and measurement are based on the procedure known in the literature as the polar check (10.17). This will now be illustrated using the example of a three-dimensional workpiece (Fig. 10.25).

The coordinates of the surface centre of gravity are obtained in the first scan of the picture. In a second scan one or more circles are made electronically around the computed centre of gravity with radii selected in such a way that they intersect the workpieces at significant points. The intersection points of the circle with the workpiece are determined in a digital circuit. This is done by comparing the circle equation $(x-x_F)^2 + (y-y_F)^2$ with centre at the centre of gravity (x_F, y_F) with the values $(R + \Delta R)^2$ and $(R - \Delta R^2)$ for each picture point. ΔR is a permissible tolerance for the radius of the circle. This tolerance margin is necessary for processing the grid points in the picture.

If the circle condition:

$$(R - \Delta R)^2 < (x - x_F)^2 + (y - y_F)^2 < (R + \Delta R)^2 \qquad (10.6)$$

is satisfied for a picture point and at the same time the level of the video signal changes then the coordinates of this point are stored in memory as the intersection point of the workpiece with the circle.

Fig. 10.26 shows the binarised and grid monitor picture for one workpiece orientation with the electronically superimposed centre of gravity and circle based on it. By joining the point of intersection to the centre of gravity an angle sequence $\alpha_{i,a}$ is obtained which is characteristic for this orientation a. The index i identifies the individual angles in anticlockwise order. The angle $\alpha_{o,a}$ gives the orientation of the angle sequence with respect to the system of coordinates.

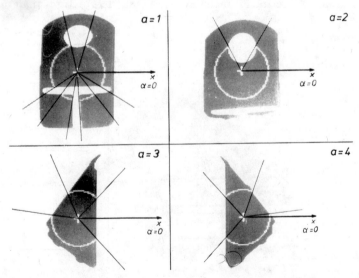

Fig. 10.27 *Possible orientations of the workpiece with the associated angle sequences (10.15).*

This angle sequence together with the already stored area forms a feature vector c_a which is characteristic for the orientation class a. If this is not the case by choosing additional radii other feature vector components $\beta_{i,a}$, $\gamma_{i,a}$ are formed until it is possible to make a final separation between the classes.

Fig. 10.27 shows the stored angle sequences for each of the 4 possible workpiece orientations.

During the learning phase all objects are presented to the senior in each of the orientations. A feature vector is formed and stored for each orientation of the object.

The features which are used are:

– the area F,
– the number P_t, $(t = 1, \ldots, T)$ of the intersection of the workpiece with the circle of radius R_t (T is the number of radii used),
– the angle sequences $\alpha, \beta, \ldots$ for the different circles.

These are then used to form a feature vector c_a serving as a reference for each orientation a:

$$c_a = (F_a, P_{t,a}, \alpha_{i,a}, \beta_{i,a} \ldots).$$

Here the angle with the index i = O (e.g. B. $\alpha_{o,a}$ in Fig. 10.26) is not a vector component. It is not invariant with respect to orientation and serves as on orientation feature for calculating the angle of twist.

During the measuring process one of the orientation classes input during the learning phase is presented to the sensor in any position. By comparing the feature vector formed during the measuring process with all learnt reference vectors the correct orientation class of the workpiece is selected by the nearest neighbour rule.

The discrimination process involves sequential comparison of each of the measured features with the stored reference features. Beginning with the measured area the number of possible pattern classes for the feature vector being classified is reduced. A rejection is made if there is no agreement on the area. A subsequent comparison of the number of intersection points further reduces the number of possible orientation classes. The remaining orientation classes are separated by comparing the measured angle sequences with the stored angle sequences. The angle sequences are rotated towards each other in the computer until there is maximum coincidence.

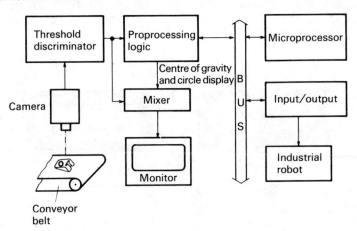

Fig. 10.28   Block diagram of the sensor system (10.16).

The rotation required determines at the same time by how much the angle of the object being classified is rotated with respect to the same angle obtained during the learning phase. In this way the rotation of the measured workpiece can be calculated with respect to a stored reference orientation.

Orientation class, coordinates of the centre of gravity and the angle of twist of the workpiece about the centre of gravity are thus determined and completely describe the position of the workpiece.

The construction of the sensor can be seen from the block diagram shown in Fig. 10.28.

The workpiece is scanned by a television camera and the picture binarised in a subsequent threshold discriminator. The numerator and denominator of the centre of gravity coordinates are formed in the following preprocessing logic during the first picture scan. The result reaches the microprocessor via the data bus. The microprocessor performs the division and stores the area and coordinates.

During the second picture scan a circle is made based on the centre of gravity. Then for every point the comparison of equation (10.6) is made. If the condition is satisfied the video signal for this point is mixed into the monitor picture. At the same time the intersection points of the circle with the workpiece are determined in the preprocessing logic and stored in the microprocessor. Other circles of different radii are made for successive scans.

The essential units of the preprocessing logic are measuring counter, multiplier, adder and register. The interconnection of these modules by means of a data multiplexer allows different computing operations to be performed.

When the measuring process has been completed the coordinates of the intersection point are converted to an angle sequence in the microprocessor system and the classification carried out. The position data of the workpiece (orientation class, centre of gravity coordinates, angles of twist) are displayed in the input/output device and communicated to the manipulation system.

As the previous systems, this sensor implementation also suggests itself for use in gravity control where the absence of features can be checked with programmed circles.

The following data give an idea of the performance of the sensor. For on-line processing during the television scanning approximately 200 ms per grid point are needed with conventional TTL digital logic. This only allows the evaluation of every second television line which together with the definition of the picture frame means a resolution of 175 points in the x-direction and 130 points in the y-direction.

The angular resolution determined from the ratio of the size of the workpiece to the picture field is about 1°.

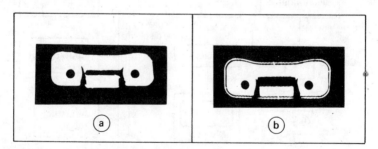

Fig. 10.29   Binary images in direct light.
(a) Front of the workpiece.
(b) Back of the workpiece.

140

The measuring and processing time is about 1/4s. The memory requirement is of the order of 4K bytes for a word length of 8 bit. The feature vector of a workpiece requires a memory capacity of about 0.1 K byte.

The illumination of the object scene plays a key role for the entire picture evaluation. Both back lighting as well as direct lighting are possible. Even though according to the results of the previous section considerable soiling of a transparent conveyor belt does not interfere given a suitable threshold of the binarising stage direct lighting is often preferable in industry because of its universal applicability.

Good results are obtained with a square illumination field with sides about 1 m long at a distance of 1 m from the picture field (10.18). About six fluorescent tubes are arranged in parallel to form the luminous field. Combinations of ring-shaped concentric fluorescent tubes and rod-shaped tubes are also suitable (10.19). The scene is processed by the television camera through an opening in the middle of the illuminating equipment. The illumination output lies between 300 W and 400 W. The conveyor belt is preferably made of black material.

Fig 10.29 shows the binary picture of the front and back of a workpiece with direct lighting. The front and rear sides can be clearly distinguished by the different shadow projections at the convexities and edges. This shadow projection is independent of orientation with the previously described illumination system. The generation of an orientation-independent and defined shadow projection or corresponding reflection phenomena is the most important condition for reliable picture evaluation using direct lighting. With the workpiece images produced by direct lighting the same features are evaluated as with contour images.

Many workpieces whose contours cannot be discriminated or only with difficulty for different orientations on a transparent conveyor belt cannot be classified using back lighting. Direct lighting opens up many possibilities here. However it must be ensured that the resulting image is not changed in an undefined manner from workpiece to workpiece by soiling (rust, scaling, oil) or reflections (gloss on ground surfaces).

10.1.7 Television sensor for orientation and position determination of freely suspended workpieces

In the previously described tasks for workpiece recognition the orientation of the objects is fixed in one plane (conveyor belt). This no longer applies to the practically important case of a suspended workpiece. The workpiece must be removed from the suspension gear by hand or with lifting devices, resuspended or transported to production equipment. In this case there are many possible orientation classes of the workpiece so that a 'mask method' whereby learnt patterns serve as masks cannot be applied.

The following sensor system is more elaborate than the previous systems because of the complexity of the task. The requirements of this task bring us to the limit of what is feasible today with microcomputers in sensor tasks. With more difficult tasks it is necessary to use a minicomputer.

The following features are used for this task: area F, centre of gravity x_F, y_F and principal moments of inertia I_1, I_2 of the silhouette (10.20, 10.21). The relationship between an object in space and its silhouette can be seen from Fig. 10.30. The object is referred to a suitable coordinate system (I, II, III), e.g. the symmetry axes of the workpiece. For a known distance the orientation of the object relative to the optical axis of a sensor is given by the angles α, β, γ.

Fig. 10.30 shows that the shape of the silhouette is a continuous function of the two orientation angles α and β. A comparison with discrete values input during a

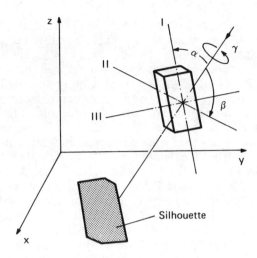

Fig. 10.30   Object and silhouette (10.21).

learning phase (mask comparison) is not feasible because of the multitude of orientation possibilities.

The principle of the sensor system involves setting up an internal model of the object. This takes place by storing a table for the functions $F(\alpha,\beta)$, I, (α,β), and $I_2(\alpha,\beta)$ for a sufficient number of orientations $x_j(\alpha,\beta)$. The corresponding values F, I_1, I_2 are then calculated during the measuring phase. By comparing with the values in the table the corresponding angle pair α,β can be found.

There are two different methods which suggest themselves for this comparison: nearest neighbour classification and sequential classification. With a nearest neighbour classification for an unknown orientation x_j with the variable $m^F(x_j)$, $m^I(xj)$, $m^{I2}(xj)$ and orientation x_i is sought for which the distance function $e(x_i, x_j)$ has a minimum (10.21):

$$e(x_i,x_j) = \sum_{K=F,I_1,I_2} \left[m^K(x_i) - m^K(x_j)\right]^2 = min. \qquad (10.8)$$

Because of the many orientation possibilities the tables can be very extensive. Since all data have to be processed before a discrimination is made this approach takes up considerable computing power.

The sequential classification is less costly in terms of processing time as will be illustrated by the following example. Fig. 10.31 shows a series of silhouettes of a workpiece rotated in steps of 30 degrees about a possible axis.

The relationship given in Section 8.2.2.1 apply to the area F, the centre of gravity x_F, y_F as well as the principal moments of inertia I_1, I_2. Fig. 10.32 shows the curves of the functions F, I_1, I_2 of the workpiece in question for a rotational angle range of $0°-180°$.

When classifying an unknown orientation, values must be obtained from these curves closest to the features of the unknown orientation. Instead of interrogating all features at the same time they are assessed one after the other. In the learning phase the tables are already ordered with increasing values so that some of the processing effort is shifted to the learning phase.

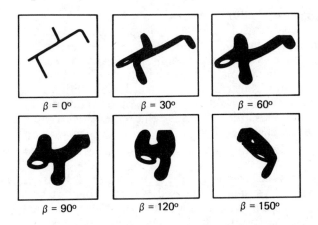

| $\beta = 0°$ | $\beta = 30°$ | $\beta = 60°$ |
| $\beta = 90°$ | $\beta = 120°$ | $\beta = 150°$ |

Fig. 10.31   Binary pictures of discrete orientations of a workpiece (10.21).

The order in which the features are processed is given by their discrimination power which can be determined by histogram analysis as discussed in Section 8.3.4.

In this case the features are effective in the order F, I_1, I_2. The meeting accuracy and a safety margin determine the search intervals (F_{min}, F_{max}), (I_{1min}, I_{1max}) and (I_{2min}, I_{2max}). Beginning with F, sublists are sought in the lists with values F_i or I_{1i}, I_{2i} as the case may be all lying within the search interval. The search is continued with the obtained sublist until an unambiguous value pair α, β is obtained. If even after all three features have been processed there is still no unambiguous result more features must be brought in or the workpiece rejected.

This table search establishes the values of the angles α, β; the angle γ about the optical axis is determined by calculating the angle between I_1 and the y-axis and a further table comparison. The orientation of the workpiece is then defined by the angles α, β, γ. The implementation of the system is similar to that described in the previous sections.

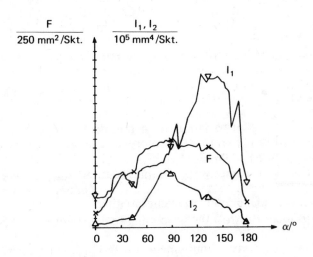

10.32   Examples for the functions F, I_1, I_2 (10.21).

143

10.1.8 Modular recognition system

With the sensor systems described previously solutions are obtained which are tailored to the specific task.

A further step is to develop different hardware and software modules which can be combined in a suitable manner for a new task only leaving adaption work to be carried out. This modular structure which is flexible with respect to different tasks represents the most advanced sensor design for industrial applications at the present time. The versatility which is achieved enables a rapid expansion of the range of television sensor applications which has been missing up till now.

The basic structure of a modular sensor system will now be described (10.22, 10.23).

Fig. 10.33 shows the structure of the system when it is fully expanded. In many applications a smaller system can be used depending on the task.

The signal provided by a television camera or a video recorder for test purposes is binarised in the signal preprocessing stage by means of threshold. At the same time the synchronising pulses from the picture are separated. All system components communicate with each other via a central bus system.

The memory modules can be divided into two groups according to their function: a RAM for storing variable data and a ROM for storing fixed data.

The RAM performs the following functions. As a contour memory it receives the coordinates of the edge points of the binary picture. When being used as a template memory the context is the comparison template. The template is computed during the teach-in phase by the microprocessor and stored in memory or transferred from the template library. The intersection points of the object with a template (e.g. circles) are determined on-line and the results loaded into a memory module designated as the results memory. An output memory is provided for displaying the picture data on a monitor.

The components falling under the heading of the library in Fig. 10.33 are ROMs and PROMs which are used for setting up programs, templates and feature vectors.

Depending on the expansion level of the system the program contains routines for sequence control, feature extraction, classification, etc.

The templates are kept in the template library so that they are ready for any application.

The use of a data library a rapid change-over with frequently reoccurring work-pieces by exchanging the feature vectors stored in the PROMs. Because of the compressed representation of the binary picture (point coordinates of changes in brightness stored in one line) edges which run parallel to a television line are not stored as a sequence of points. However, since the enclosed edge is needed for determining the points of intersection between the object and the template it is generated from the content of the contour memory by the contour line generation module.

The template shift also forms part of the memory data preprocessing. It shifts the templates contained in the template memory to a reference point of the binary picture (e.g. the centre of gravity).

Already during picture scanning simple features such as area, points of maximum extension, the numerators for the centre of gravity coordinates, etc. can be extracted by means of modular electronic circuits.

The microprocessor controls the processing sequence or processes the picture data itself.

With programs with heavy requirements on the processor, in particular with frequent multiplications, divisions and higher mathematical functions the pro-

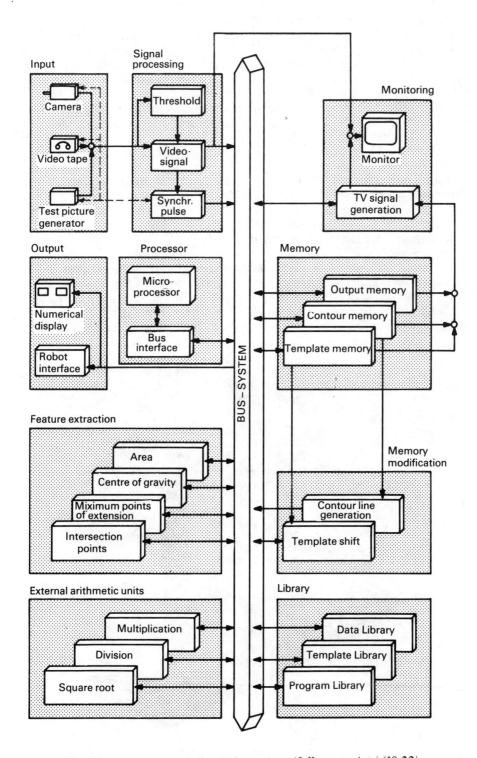

Fig. 10.33 Block diagram of the modular system (full expansion) (10.22).

cessing time can be considerably reduced by using external arithmetic units. This has been already used in the sensor systems of Sections 10.1.2 and 10.1.6.

The output of the results calculated by the microprocessor (orientation coordinates, object) can be communicated via an interface to the manipulation system as well as to a display for the operator. This can also be used to display test and diagnostic data.

The programs also have a modular structure which facilitates adaptation to new system configurations. In addition to the features determined directly be electronic circuits other features such as the moments of inertia, circumscribing rectangles, etc. are extracted by special computing programs. There are also program packets for the individual classification methods.

Up until now it has been impossible to use pattern recognition sensor systems on a large scale because after lengthy development a solution was obtained for a single tack. The modular principle allows on the other hand a very rapid expansion of the range of applications and so represents a significant contribution to broadening the scope of applying recognition systems in industry.

10.1.9   Object recognition by means of diode arrays integrated in grippers

For a large number of tasks the position of the workpiece is known approximately. However for automatic handling the position must be known with an accuracy of 1 mm (10.24). For example a task which often occurs in industrial practice is the loading and unloading of pallets: the approximate position of the object is known

Fig. 10.34   Diode array in the gripper of a manipulation system (10.25) (Source: Inst. for machine tools and production engineering TU Berlin).

but within this tolerance the exact position and orientation of the object is unknown.

For an imaging sensor which with a typical resolution of 1% of the picture field (Section 4.3.1) has to process a pallet of e.g. 1 m^2 the accuracy of the measured workpiece position is of the order of 1 cm and is, therefore, inadequate. By reducing the picture field to 10 cm the accuracy would reach the required value of 1 mm. The limitation of the picture field is achieved if the imaging sensor is incorporated in the gripper of the handling system. The handling system automatically moves to the known and preprogrammed object position and the exact measurement of the position is now carried out in a restricted picture area.

Because of their small measurements diode lines and arrays suggest themselves for incorporation in grippers (Section 4.1.3).

Fig. 10.34 shows an arrangement of diode arrays integrated in the gripper of a manipulation system (10.25). The system works with two image transducers for position measurement. The imaging sensors used are self-scanning photodiode arrays with 1024 picture point elements.

The processing of the sensor signals for determining the fine position takes place in a similar manner to the example previously described with high-speed digital registers or by means of a microprocessor system. The choice of suitable illumination is also especially important here. Back lighting cannot be used for example for the position recognition of palletised workpieces. With direct lighting recognition is made difficult because the presence of the gripper over the object makes it impossible to use the large-area homogenous lighting described in Section 10.1.6. This produces reflection images superimposed on the object. By using a lighting bar which as can be seen from Fig. 10.34 generates an approximately homogenous illumination density parallel to the pallet surface these disadvantages can be overcome to a large extent.

10.1.10 Television sensor for image-assisted welding

Up until now there has hardly been any application of sensors for controlling welding processes although the preconditions are already present in mechanised welding installations (10.26). The reason for this is that the information on the welding process has only been derived up until now from the arc or temperature parameters at points with only limited significance. On the other hand the picture of the welding point contains so much significant information that the welder can use it to carry out the welding process by varying the parameters such as material feed and torch manipulation. A promising approach is, therefore, to control the welding process using the picture of the welding point.

When developing an optical sensor for processing the picture of a welding point extreme differences in brightness have to be taken into account. Humans have the capacity to process images of this type and to assign different picture contents to objects. A welder is not disturbed by the presence of the arc when observing the darker welding point. He can imagine the parts of the welding site which are obscured by the arc plasma. This performance cannot be equalled with a television camera and subsequent electronic evaluation circuit. Kept within reasonable cost limits, the interfering effect of the arc must, therefore, be overcome by suitable measures.

A television sensor will now be presented which for TIG root fusion determines a regulating variable from the picture of the welding point and enables the feed of the filler to be controlled (10.26, 10.27).

The welding process is illustrated in Fig. 10.35. The edges of the workpiece are fused together by an oscillating forwards movement of the torch. A ring-shaped

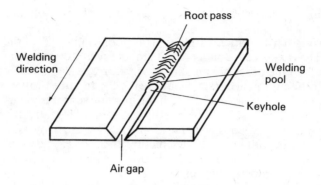

Fig. 10.35 Illustration of the welding process (10.27).

extension of the air gap is formed, the keyhole. As soon as it has exceeded a certain size the welder guides the wire into the welding pool. This introduction of the welding wire can be automated if the shape of the keyhole can be ascertained. Investigations have shown (10.26) that even with asymmetrically distorted key-holes (Fig. 10.36) the chord length in the y direction as a function of position x (welding direction) corresponds to an ideal circle to a large extent. The diameter and the x area of the circle corresponding to the chord profile are then according to Fig. 10.37 features for the guiding of the welding wire.

A television camera suggests itself for processing the keyhole. Because of the strong overexposure only semiconductor cameras can be used (Section 4.3.7). In addition their high sensitivity in the red and infra-red spectral regions is exploited (Section 4.3.6).

The welding current is periodically reduced or completely cut off for the picture scanning. Then the glowing welding point illuminates itself in its own light without the interference of the arc. The optical path from the welding point to the television camera is only made free during these time intervals by means of a synchronised rotating aperture disc. This brief exposure is sufficient to generate a charge image which is scanned according to the usual television standard.

By means of a threshold the television picture generates a binary picture in which the keyhole appears as a dark area. This area is ascertained by a line-by-line

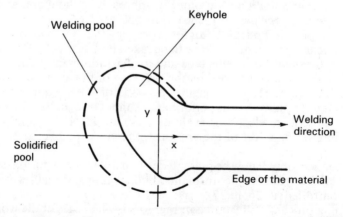

Fig. 10.36 Diagram of a keyhole (10.26).

148

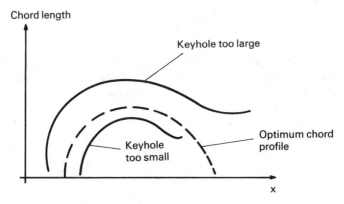

Fig. 10.37   Keyhole and chord profile.

measurement of the chords and summing in a start-stop counter. When this variable exceeds a limit value the welding wire feed is released.

10.1.11   Image sensor with sequential classifier

A recognition procedure for workpieces will now be presented as an example of a sensor with a sequential classifier (Section 8.3.4) (10.28). The parts lying on the conveyor belt are imaged by the television camera and converted to a contour image by a computer (Fig. 10.38).

The following features are obtained from this image:

c_1 = length of the contour
c_2 = area
c_3 = area of the holes
c_4 = minimum circumscribing radius
c_5 = maximum circumscribing radius

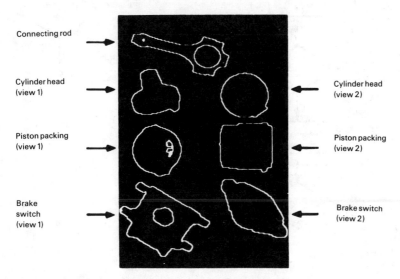

Fig. 10.38   Contour images of various workpieces (10.28).

149

A histogram analysis (Section 8.3.4) is made with respect to these features for each of the workpiece classes shown in Fig. 10.38. This leads to the decision tree shown in Fig. 10.39.

In this example with sequential interrogation, an average 2.7 features have to be processed until the final classification is made. The processing effort compared with a nearest neighbour classifier which always needs features for discrimination is, therefore, considerably small.

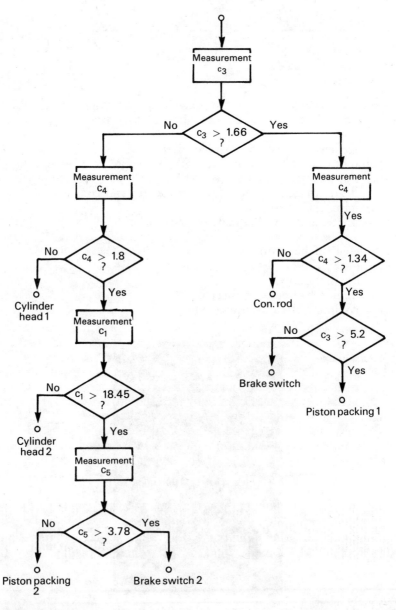

Fig. 10.39 Decision tree.

10.1.12 Television system for the visual inspection of conductor patterns:

The sensor systems presented in the previous sections are just as suitable for quality control tasks as for object identification. The optical correlator (Section 10.1.1) can be used to inspect the overall pattern if the defect represents a considerable proportion of the surface of the pattern under investigation (integral method). The television systems of Sections 10.1.5 and 10.1.6 can always be used for inspection tasks if a defect lies within a certain known geometrical tolerance. The range covered by this tolerance can then be monitored by suitable positioning of the corresponding scanning lines and circles.

However, there is a large number of tasks in visual inspection where it is necessary to search entire areas for tiny defects not previously localised. An example of this is the inspection of printed circuits. In this case fine cracks, notches or holes can occur at any place. When the patterns have small dimensions, e.g. with conductor patterns for the assembly of integrated circuits the inspection is carried out with the help of microscopes.

It is understandable that humans carrying out such tasks where a large number of items are involved are under considerable stress because of the monotony of the work and the high level of concentration required at the same time. Despite the checking which is carried out, this is the reason for a high reject rate through unrecognised defects.

A television system will now be presented for the automatic visual inspection of conductor patterns for integrated circuits which is capable of very rapid detection of tiny defects on small patterns (10.29).

Fig. 10.40 shows a section of a strip of foil with the connecting paths for assembling integrated circuits. The dimensions of each of the three patterns shown in Fig. 10.40 are about 20×8.2 mm. The narrow paths in the middle on which the integrated circuit is bonded (bonding zone) are only 0.1 mm wide. A crack is shown in the centre pattern. A marker is punched to indicate the defect.

In order to inspect the conductor paths the foil is scanned by two television cameras which differ in their accuracy of processing different areas. The pattern is divided into four fields as shown in Fig. 10.41. Fields 1, 2 and 4 are imaged by the same camera and together make up the overall pattern. Field 3 which is imaged by another camera with a stronger magnification only involves the bonding zone which has higher accuracy requirements. The strip is observed with back lighting. The transparent foil produces a good contrast for the conductor paths.

During inspection the width is checked for each conductor path and for each insulation strip between two conductors. In this connection each conductor pattern to be tested is compared with a model (reference pattern) in which the centre line of these conductor paths as well as their minimum required widths are laid down. The reference pattern with which a pattern under test is compared consists of the centre lines of all conductor paths and insulation strips of the pattern. For each of these

Defect in the conductor path

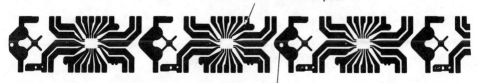

Marking of the defect

Fig. 10.40   Test patterns of the conductor paths (10.29).

151

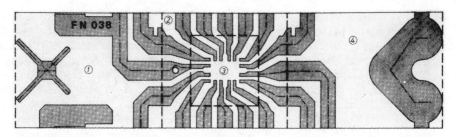

Fig. 10.41   Picture fields for the two television cameras (10.29). (Source: Philips).

centre lines the minimum required width of the path is stored in memory as well as indication as to whether it is a conductor or insulation. Corresponding to the four picture areas there are four reference patterns for each pattern.

The required course of the centre lines and widths is known as a *priori* and is input into a computer by paper tape for each reference pattern. When inspecting a conductor path, it is checked for each centre line point whether the measured test pattern contains an adequate number of conductor path points symmetric to this point. This check is carried out in a horizontal or vertical cross-section of the path depending on the definition of the path width in the reference pattern.

In this procedure the exact relative positioning of the stored reference pattern and the measured test pattern is important. For this the position of the test pattern is determined and the reference pattern moved until both patterns coincide.

The position of the test pattern is determined by imaging and evaluating two significant points in two picture windows B_1 and B_2 (Fig. 10.42). The arrangement of the picture windows is selected in such a way that when the reference model is formed a vertical edge of a conductor path lies in the middle of one window and a horizontal edge of a conductor path lies in the middle of the other window. The values x_m and y_m then define the position of the reference pattern. When an image has been made of a conductor pattern and transferred to memory the position of the conductor path edges is determined in these windows. The difference of these measured values to x_m and y_m then determines by how much the reference pattern must be displaced to ensure a coincidence of the centre lines of the reference pattern and the conductor pattern under test. When inspecting the conductor paths tiny measurements have to be processed. The inspection of the picture field 3 in Fig. 10.41 requires a resolving power of 10 μm. The camera lens is chosen so that the

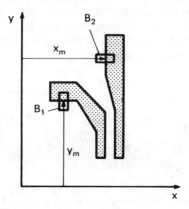

Fig. 10.42   Measurement of the position of the test pattern.

152

distance between two picture points correspond to 10µm in the pattern. An industrial television camera with a resolution of about 1% of the picture field (Section 4.3.1.) can only image a picture field of 1 × 1 mm. This is considerably smaller than the picture field to be inspected. Therefore, a modified camera system with an enhanced resolution is used so that with a frame of 400 lines 450 picture points can be processed. This camera then covers the required picture field of 4 × 4.5 mm. On the other hand a resolution of 20 µm is adequate for the other picture fields 1, 2 and 3.

In Fig. 10.43 a vertical conductor is shown by means of crosses and the insulation by means of dots. In order to increase the reliability of recognising edge points of the conductor a certain number of points is fixed as a criterion. If points are investigated symmetrically positioned to the centre line of the conductor cross-section then the following situation can arise:

(a) All points are present,
(b) Only one edge point is missing,
(c) One or more points are missing inside the conductor.

In the first case the cross-section is in order. In the second case there could be a positioning defect between the test pattern and the reference pattern. This effect can be overcome by repeating the symmetry test of the cross-section with the centre line displaced. In the latter case there will be a recognisable defect present.

```
· · x x x ✳ x x x · ·
· · x x x ✳ x x x · ·
· · x x x ✳ x x x · ·
· · x x x ✳ x x x · ·
· · x x x ✳ x x x · ·
· · x x x ✳ x x x · ·
```

Fig. 10.43   Stored image of a conductor (10.29).

As soon as a defect has been found in one conductor, an indication is made in a projection register PB as to which points are missing for all subsequent cross-section. Two points on either side of the conductor are also examined. If the defect involves a hole in a vertical or horizontal conductor a projection of this hole is made on to a perpendicular cross-section of the conductor (Fig. 10.44). The maximum width of the hole (here four points) as well as the minimum width of the free passages adjacent to the hole (here one and two points) can be now read off from the content of the projection register. Thus the projection register stores a projected image of the defect. The necessity for this will be illustrated by the highly

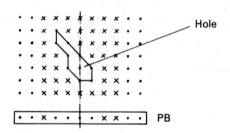

Fig. 10.44   Detection of two-dimensional defects (10.29).

153

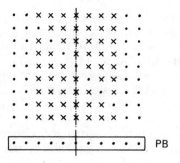

Fig. 10.45   Detection of a scratch (10.29).

interesting practical case shown in Fig. 10.45 of a narrow scratch on the conductor. There is only one defect present in every conductor which in itself is no reason for rejection. It is the projection which indicates that the connection is interrupted.

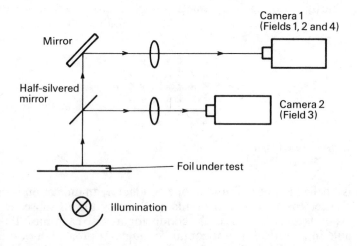

Fig. 10.46   Picture scanning for foil inspection.

Fig. 10.46 shows a diagram of the imaging device of the system. By means of mirrors and various lens arrangements camera 1 scans a picture field of 4×4.5 mm with a resolution of 10 µm. The foil is moved in front of the camera by means of a stepping motor so that the individual areas come into the picture field. The video signals coming from the cameras are alternatively routed via a multiplexer for picture processing. In this way the picture is input on-line to the minicomputer carrying out the comparison via two shift registers, each of which can handle the information of one line (450 bits).

The complete testing cycle lasts about 1 second for one pattern and is, therefore, faster than a human operator. At the same time this kind of automatic system is capable of providing statistical data on the defects which have been found. In this way certain manufacturing faults can be revealed which otherwise would be difficult to recognise e.g. a slight corrugation of the foils, etc.

10.1.13 Visual inspection tasks with grey-tone processing

With a number of industrial tasks it is not possible to reduce the picture to a binary image in such a way that only the objects of interest are available for processing. Frequently reflections, soiling or the background of the object generate undefined images varying from object to object. The binary processing method presented in Section 10.1.6 whereby a defined object is generated under direct lighting does not then apply.

An example will now be described of applying grey-tone processing to such cases (10.30). Here the position is checked and a visual inspection is made of a transistor semi-conductor chip on a base (Fig. 10.47).

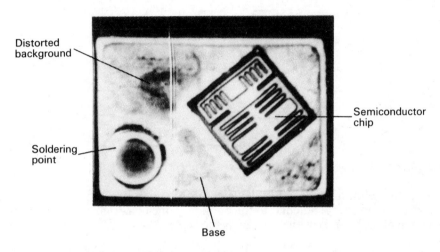

Fig. 10.47   Object under examination (10.30).

The task consists of checking the position and orientation of the semi-conductor chip whereby the semi-conductor chip must not be twisted too much, not be too near to the soldering points and not protrude beyond the confines of the base. At the same time it is checked whether bits have not broken off from the semi-conductor, whether it has been cut too small or whether it is missing altogether. With the orientation information obtained it is possible to position test peaks automatically and so test the electrical functioning.

The recognition task then can be broken down into the following four steps:

(1) Forming a grid (matrix). Here 50 × 50 picture points turn out to be adequate (diode array).
(2) Determination of the orientation of the chip.
(3) Determination of the corner points of the chip. This fixes the exact position.
(4) Testing the chip for completeness.

The orientation of the chip is determined using the gradient methods discussed in Section 8.2.2.3.

In Fig. 10.48 the picture being investigated with brightness E is surrounded by 8 picture points with brightness A, B, C, D, F, G, H, I. According to equation (8.74) the gradient can be calculated from:

$$g = a \cdot e_x + b \cdot e_y \qquad (10.9)$$

155

with

$$a = A + 2B + C - (G + 2H + I)$$
$$b = C + 2F + I - (A + 2D + I) \qquad (10.10)$$

In this case the values in the vertical and horizontal vicinity of the picture point are more strongly weighted (Section 8.2.2.3). The direction α of the gradients can be obtained from:

$$\alpha = \arctan a/b \qquad (10.12)$$

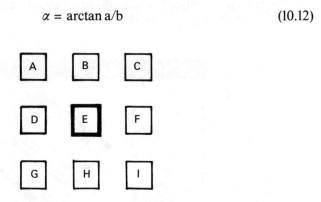

Fig. 10.48   Determination of the change in the grey scale.

Since the definition the gradient is a measure of the strongest change in brightness and is, therefore, perpendicular to the boundary line of a brightness transition, the direction of the boundary line is given by the angle $\alpha' = \alpha + \frac{\pi}{2}$.

With rectilinear objects such as often occur in industry a simple direction coding method (10.31) can be used to determine the orientation of the object even in the presence of interfering points. Even when the picture is distorted there will usually be an accumulation of picture points with especially large changes in brightness along the boundary lines. Picture points on straight lines have the same angle α or α'.

With a right-angled object, then according to Fig. 10.49, there is an accumulation of four angles each displaced by 90° to the other $\alpha' + \frac{\pi}{2}i$ (i=0, 1, 2, 3). The orientation α_0 of the rectangle can be determined from this. This simple analysis also makes it possible to detect specific objects even in the presence of image distortions. The process can be repeated in a second iteration step where in order to enhance the

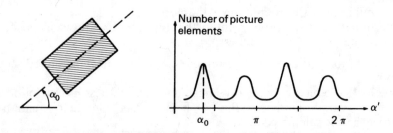

Fig. 10.49   Direction selection method

156

reliability only points are still considred which are in the immediate vicinity of the angle α_0 obtained.

It has been assumed up until now that the optical axis of the imaging camera is exactly vertically over the object. However, in many applications this is not feasible in practice. If the object is viewed from a point sufficiently far away with an angle γ between the normals to the object plane and the optical axis of the camera an altered angle sequence βi is produced from the angles $\alpha' + \frac{\pi}{2}i$ (1 = 0, 1, 2, 3) Fig. 10.50). The following relationship applies to this case (10.31).

$$\tan\left(\alpha' + \frac{\pi}{2}i\right) = \tan\beta_i/\cos\gamma\,(i = 0, 1, 2, 3). \tag{10.13}$$

The 90° periodicity pertaining in the case $\gamma=0$ no longer applies now.

After the orientation of the chip has been established the next thing is to determine the four corner points which then fix the position.

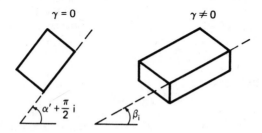

Fig. 10.50   Oblique viewing of objects.

A corner point is distinguished by the fact that a dark picture element is surrounded by bright zones in two perpendicular directions. A template comparison suggests itself as a method of finding such a prominent point (Section 8.2.2.2).

With a rectangle there are four points in the four quadrants which following geographical custom are decribed by the directions North-east, North-west, South-west and South-east. For instance the mask has the appearance shown in Fig. 10.51 for the north-western corner: The object zone (A B C D E) is surrounded by a bright zone (F G H M N P Q). In this detailed recognition a so-called local mask is used which looks for details of the object. The appropriate mask is applied to each picture in the relevant quadrant and using a quality criterion a measure is obtained of the probability that a corner point is present.

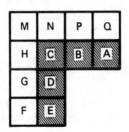

Fig. 10.51   Mask for a picture corner (10.31).

In this case the following quality criterion has proved itself (10.31):

$$G_1 = (F + 2G + 2H + M + 2N + 2P + Q) \\ - (3A + 2B + C + 2D - 3E) \qquad (10.14)$$

Corresponding relationships can be obtained for the other three corners. This quality measure has a maximum value for a corner point: the positive sum is large because of the high light intensity and the second sum is small. Depending on the task and the reflection conditions the light intensities can also be distributed in entirely the opposite way. When calculating G_1 the picture point being investigated occupies the position of the point C in Fig. 10.51.

It can be shown in this example that it is not necessary to apply the template in rotated orientations to compensate for a rotation of the chip.

A further test is made to ensure that these are really corner points. In this case, use is made of the fact that for a corner point there must be three other points which are at defined distances from one another and have an orientation angle equal to that already obtained for the orientation of the chip. A quality criterion is also defined for this so that taking into account the above-mentioned secondary conditions the four corner points of the chip can be determined. The expression:

$$G_2 = \sum_{e=1}^{4} G_1(x_e, y_e) \qquad (10.15)$$

is formed for all corner points (x,y) found as suitable by the quality criterion G_1 which, in addition, can be considered in terms of distance relationships and angles. The corner points of the chip are those four points for which the quality criterion G_2 has a maximum value. Knowing the corner points the position of the chip is defined.

The quality criterion G_2 is a global template for determining larger geometrical relationships.

The still outstanding check for completeness is made by a simple contrast measurement along the outside now known from the corner points. For this the differences in brightness between the base background and the boundary of the chip are formed. If this value falls below a certain threshold then the chip is defective at this point.

In its smallest version the system requires a minicomputer with 8 K memory. The total testing time is less than 1 second. With a picture field about 2 mm × 5 mm (double the chip size) an accuracy of approximately ± 0.1 is obtained for a resolution of 50 × 50 picture points. The angular error is approximately ± 1.2 degrees.

10.1.4 Multisensor-controlled assembly systems – a look into the future

In industrial manufacturing processes humans carry out the assembly of parts as a matter of routine. Only when an analysis is made of the assembly process during the course of automation, is it realised how varied and complicated the actual operations are. This applies especially to the requirements for the sensor systems and the kinematics of the manipulation system. Assembly is usually carried out using both arms as well as coordinated monitoring by the eyes and the tactile sense of the hands. The most complex sensor problems have to be overcome here such as the processing of grey-tone pictures from different aspects. It is, therefore, hardly surprising that up until now the automation of complex assembly tasks on a world-wide scale has only been put into practice within the framework of research projects

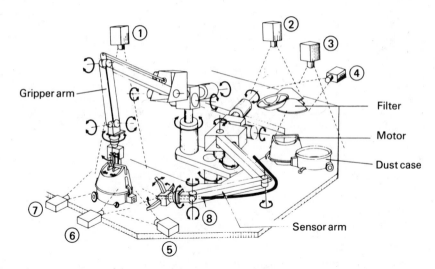

Fig. 10.52   Design of the assembly system (10.32).

in a few laboratory models. At present there are still no economically viable solutions.

We shall now present a multisensor-controlled assembly system which represents the limit of what is currently feasible (10.32, 10.33).

This example has been intentionally chosen at the final section for recognising sensor systems. As far as optical sensors are concerened it combines all the difficulties and possibilities of the individual systems discussed in previous sections ranging from grey-tone picture processing to the integrated television camera in the gripper. Although the cost prohibits any economically viable application at present, the example shows nevertheless that complicated sensor problems can be solved.

The function of the assembly system is to assemble the motor casing, the filter element and the dust case of a domestic vacuum cleaner. The motor casing is made of plastic and is fastened to the dust case by means of two clips. Before this assembly the dust case has to be fitted with a filter element which consists of a plastic frame with rubber seal and fabric lining. Because the plastic parts can be easily deformed and cause strong reflection effects, the assembly operations make heavy sensory demands and are close to practice. The individual manipulation operations are as follows: gripping the filter, transferring the filter to the assembly hand, gripping the dust case, joining the motor and dust cases, closing the sealing clips.

Fig. 10.52 shows a diagram of the assembly system. Following the human system it has two arms with 8 degrees of freedom arranged on a common base place. One arm is designed as a gripping arm for transporting heavy objects. The second so-called sensor arm has motor-driven fingers with 30 tactile transducers. These consist of springs which actuate microswitches when they have been pushed back a certain way. A television camera (8) is also incorporated in the gripper of this arm. Three other television cameras (1), (2) and (3) view the scene from a vertical direction and the four television cameras (4), (5), (6) and (7) from a horizontal direction. The optical and tactile sensor signals are processed by two separate minicomputers.

We shall now give a few details of the processing of the sensor signals. The filters to be asembled lie randomly and unseparated in the picture field of the television

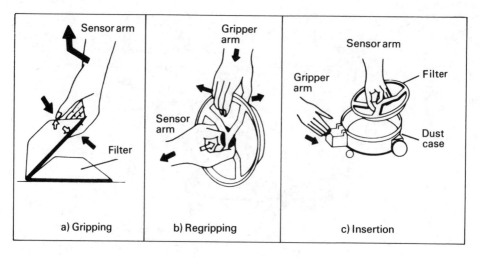

| a) Gripping | b) Regripping | c) Insertion |

Fig. 10.53   Illustration of the assembly process.

cameras (2), (3) and (4). The recognition task is to recognise the top and, therefore, uncovered filter. Here the analogue television picture showing all the filters is converted to a contour image.

After cleaning up the image to suppress interference, it is searched for completely closed oval contours of a certain size: only the top uncovered filter has a closed contour. The sensor arm is moved to the recognised filter with this rough information. A fine determination of the position of the filter edge is then carried out by means of the camera incorporated in its gripper.

With the help of the tactile transducers in the fingers the filter case is touched, gripped under the edge of the case and the filter picked up (Fig. 10.53a). In a subsequent regripping operation the sensor arm transfer the filter held at its edge to the gripper arm and picks up the filter again in a more favourable position for the following joining operation (Fig. 10.53b). This operation is controlled by the tactile sensors in the gripper of the sensor arm. Such a regripping operation occurs in every complicated assembly process in particular when two manipulation systems are interacting and is an important facility which such assembly systems must have.

The next assembly operation involves the insertion of the filter in the dust case. This takes place exclusively with the help of tactile transducers. By means of these transducers the gripper arm recognises the contact when the filter is inserted by the sensor arm (Fig. 10.53c). This signal controls the movement of the sensor arm. This feedback of pure tactile signals enables the filter to be inserted in the case by correction movements of the sensor arm.

The motor casing has next to be gripped and inserted in the dust case. The recognition of the motor casing is especially critical since this is a shiny plastic part with particularly disturbing reflection properties. By means of suitable illumination measures (Section 10.1.6) a definitive image is generated of the reflections in direct lighting and after a threshold operation the binary image of the reflection phenomena processed.

During insertion the scene is viewed alternatively by the two camera (5) and (7) at right angles to each other and first the distance between the two parts motor and dust case minimised in the viewing plane. When both positions coincide in the plane the two parts to be joined are aligned with respect to the orientation angle using the video signal from camera (1). The alignment mark is a projection for the clamping fixture which can be recognised in the grey-tone picture. When both parts have the correct position the motor held by the gripper arm is joined by a vibratory movement to the dust case held by the sensor arm. A force transducer in the gripper arm limits the joining force and terminates the joining operation when the boundary has been reached.

The last assembly operation involves repositioning the two clips for holding the two parts. This is done with the help of tactile transducers in the finger grippers. The end of the clamping operation is derived together with a force measurement from the angular position of the gripper of the sensor arm.

The entire assembly process requires 2 minutes. Even though such a system cannot find any economically viable application in industry at the present time it shows, nonetheless, that even relatively complicated sensor tasks can be solved given sufficient outlay. This example gives an idea of what is possible today. An entire range of extreme difficulties have been successfully overcome here in the area of sensor technology such as the processing of grey-tone pictures, the transfer of workpieces from one manipulation system to another, the tactile feeling of a required position and last but not least the integration of highly diverse tactile and visual sensor signals.

10.2 Acoustic sensor systems

The main tasks of acoustic sensor systems are in quality control in the production sector, in the surveillance of installations for early warning of faults and in process monitoring and control.

Since processing acoustic signals involves time-dependent or frequency-dependent signals which only have one dimension in many cases, it is possible to manage without a computer system by means of adaptive signal processing (Chapter 8.2.1). However, this does not mean that acoustic signal processing is simpler than video signal processing. Often the task is even more difficult. With a complex piece of machinery, e.g. an internal combustion engine the noise is generated by a large number of sound sources. These sound sources are superimposed on each other and are not distinguished by the measuring transducer but taken as a whole. These difficulties explain the relative rarity of acoustic pattern recognition systems in industry even on a worldwide scale. However, over the next years it can be expected that the methods of acoustic pattern recognition will be more widely adopted especially in the industrial production sector.

Some examples of acoustic sensor systems for industrial applications will now be presented.

10.2.1 Gear Testing

Rotating pairs of gears produce a noise, so-called gear crashing. During quality control the gears are usually evaluated by a human inspector by assessing the running noise. An automatic measuring system for acoustic quality control of gears will now be described (10.34, 10.35).

A detailed investigation (10.35) shows that the gear noises can be broken down into three components: short-period components which occurs with each tooth meshing, long-period components which vary in synchrony with the revolution of the gear wheel and transient components which are coupled to individual teeth.

There are three tasks for the acoustic quality control of gears:

(a) Detecting and locating locally defined defects at the gear wheel
(b) Detection of large-scale production and assembly faults (concentricity faults)
(c) Assessment of quiet running.

161

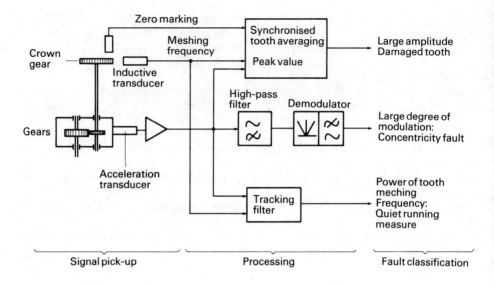

Fig. 10.54   Signal pick-up and processing for testing gears (10.35).

When making measurements in practice for assessing gears the structure-borne sound is picked up by a quartz acceleration gauge perpendicular to the surface of the gearbox. The pick-up point should be as near as possible to the noise source where the teeth mesh since the wanted signal is weakened at every mechanical separation point (Section 5.6). The most suitable coupling of the transducer is a rigid connection by screwing down. However, with production testing this cannot be considered because of the long set-up time. It then makes sense to use a magnetic clamp for the coupling. The transfer factor for this coupling mode is linear in the frequency range 0-3 kHz. The significant structure-borne sound components also fall into this range.

Methods for determining the three types of fault will now be presented. Fig. 10.54 shows the block diagram of the measuring system. There are three different processing channels corresponding to the three different types of task.

(a) Assessment of gear wheel damage

Fig. 10.55 shows the structure-borne sound signal for a damaged gear wheel. It is charcteristic here that individual transient impact pulses exceed the average signal level.

In this case it makes sense to evaluate the time variation of the structure-borne sound signal. A tooth-synchronised peak value measurement indicates the associated point on the gear wheel or the defective tooth. The reference signal for synchronisation is either taken from the gear wheel itself or picked up optically or inductively from a crown gear. Different crown gears are necessary for different gear wheels so that several types can be tested with the same arrangement. For a pair of gears two measuring arrangements should be provided in parallel for the gears concerned. The method is more reliable if the time variation of the amplitudes is synchronised and averaged over several revolutions (Section 8.2.1.2). An appropriate zero marking at the beginning of a revolution is also derived from the crown gear.

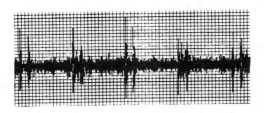

Fig. 10.55   Structure-borne sound signal for a damaged gear wheel (10.35).

(b) Assessment of concentricity faults

Concentricity faults are long-period manufacturing and assembly faults. They show up as amplitude fluctuations which can be seen in the time variation of the signal by lines joining the maxima. The amplitude modulation of the broad-band structure-borne sound (carrier) as a result of mechanical irregularities (modulation) generates two symmetrical side bands to the original carrier with the same band width which are added to the carrier (Fig. 10.56).

The power density spectrum of the demodulation signal (Fig. 10.57) is a convenient way of diagnosing concentricity faults. The spectral power distribution has several lines, one of which can be assigned to the drive frequency F_1 and another to the driven frequency F_2. The two frequencies are related by the gear ratio $G = F_1/F_2$. Thus the spectrum of Fig. 10.57d illustrates a gearbox with $G = 4.36$ with 2600 rpm at the drive, whereby the lines at 41.6 Hz and 9.5 Hz are assigned to the drive and driven wheels respectively. The line structure can be interpreted in terms of defects in both gears. A condition for spectral analysis is the presence of constant rotational speeds. If the rotational speed is variable the degree of modulation characteristic can be derived from the envelope.

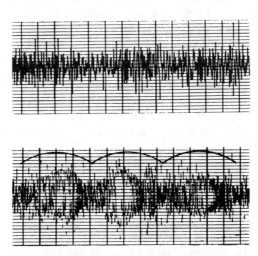

Fig. 10.56   Structure-borne sound signal with and without concentricity faults (10.35).
(a) Gears without concentricity faults
(b) Gears with concentricity faults

163

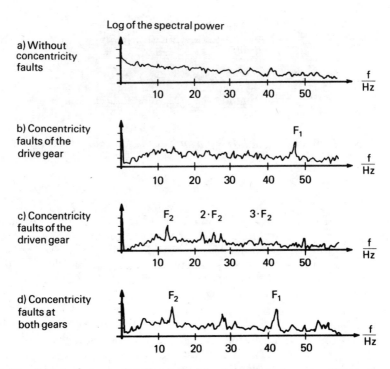

Log of the spectral power

a) Without concentricity faults

b) Concentricity faults of the drive gear

F_1

c) Concentricity faults of the driven gear

F_2 $2 \cdot F_2$ $3 \cdot F_2$

d) Concentricity faults at both gears

F_2 F_1

Fig. 10.57   Spectral power distribution of the demodulation signal (10.35).

(c) Assessment of quiet running

It can be shown that the disturbed running of the pair of gears is essentially correlated with the power of the meshing frequency (10.35). Under stationary conditions this characteristic for quality control can be taken from the spectrum. When making dynamic measurements of the characteristic with different rotational speeds it is necessary to extract the meshing frequency as a function of the rotational speed and to ignore other noise sources and their signal components. Here it makes sense to use a filter procedure based on the principle of cross-correlation (Section 8.2.1.4). The procedure (Fig. 10.54) makes the narrow-band selection of the meshing frequency from the signal and makes it possible to measure its power. The quality characteristic can be followed over the entire range of rotational speeds and for instance the maximum value stored for assessing the gears.

10.2.2   Steel converter monitoring

The monitoring of the oxidation process in steel converters described below is an example of acoustic process control (10.36, 10.37). It is not the processing of the measured values which causes problems but the extraction of a significant parameter during the analysis. At the same time in terms of the harsh environmental conditions this example shows how difficult it is to obtain the measurement signal in an industrial process.

During steel production, the accompanying elements in the crude iron (carbon, sulphur, phosphorous) are reduced to the required proportions by oxidation. This takes place in the steel converter earlier known as the Bessemer or basic converter which involves blowing in air or pure oxygen. The course of combustion in the smelt is essentially controlled by the supply of oxygen. For a long time the chief engineer used the optically observed flame picture and the converter noise as a criterion for the control. Today both assessments are made difficult by the heavier requirements on waste gas cleaning. For this it is necessary to have a hood over the converter which affects the observation of the flame and the propagation of the sound. With a steel converter the oxygen stream emerging with high velocity from a jet excites the gas mixture in the converter inner cavity to sound vibrations which emerge with the stream of waste gas via the converter mouth.

The resulting sound varies during the blowing period on the one hand because of the change in volume of the gas-filled inner cavity through slag formation and on the other hand because over time more and more liquid slag surrounds the air inlet thus affecting the sound of the emerging oxygen. The slag being formed covers the smelt with an ever increasing layer and muffles the sound emanating from there. Sound pressure changes are produced of 1:10. The frequency range of these variations can differ for different converter types.

Fig. 10.58 shows the basic arrangement for acoustic converter monitoring. The low-frequency component is filtered out from the total spectrum which contains strong high-frequency components (hissing) using the microphone signal supplied by a preamplifier. In practice the sound pressure measurements are made in a frequency range of 150 Hz – 1000 Hz. As a rule this corresponds to the most significant slag formation. The rectified signal is proportional to the variation in amplitude of the sound in the selected frequency range.

The basic variation of the sound pressure is shown in Fig. 10.59; it depends strongly on the converter type and frequency range. The strong slag formation causes a significant drop in sound pressure which can be used to derive a control signal for stopping the supply of oxygen. The exact appraisal of this instant is important so that the impurities are not reduced too much and the slag does not boil over.

The arrangement of the microphone is important for the mesurement. On the one hand the sound should have unhindered access to the microphone. On the other hand it must be protected against heat, flying sparks, slag eruptions and soiling. Following Fig. 10.58 this is accomplished by conducting the sound through a 5-15 m long steel pipe (diameter 3-5 cm) from the converter exit to the microphone (10.36). The microphone is tightly secured close to the open end of the sound conductor pipe. An alternative solution is to attach the microphone probe directly to the converter wall behind protective plates by means of special rubber-metal fittings (10.37).

10.2.3 Quality control of electric motors
A procedure for the acoustic quality control of induction motors will now be

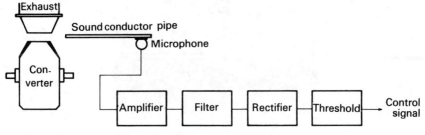

Fig. 10.58   Principle of acoustic converter monitoring

165

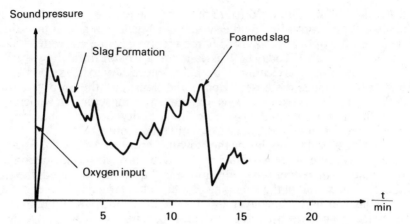

Fig. 10.59   Basic variation of the sound pressure for the steel converter.

presented as an example of a complex testing function (10.38). Manufacturing faults in small motors with a power of approximately 300 W are detected. The most frequent causes of noise are bearing defects, magnetic fields and grinding processes. Magnetic noises arise as the result of a change in the air gap between sector and rotor due to a non-centric bedding of the rotor. Grinding noises arise for instance when parts of the stator winding insulation or other foreign bodies protrude into the air gap (10.38). Bearing defects are caused by faulty bearing bushes or ball bearings.

(a) Magnetic noise
The modulation of a carrier of 800 Hz by low-frequency vibrations of 25 Hz and 50 Hz is in this case typical of a magnetic nose (Fig. 10.60). The frequency of the 800 Hz carrier can be interpreted as the harmonic of the stator current corresponding to the number of slots in the stator. The modulation is caused by an unbalanced rotor which alters the air gap or by varying geometry of the rotor slots.

The magnetic noise occurring as structure-borne sound is converted to an electrical signal via an acceleration transducer and amplifier. A level measurement using a band filter tuned to the carrier frequency of 800 Hz shows a higher level for motors with magnetic noise.

(b) Grinding noises
The appearance of high-frequency peaks in the air-borne sound signal

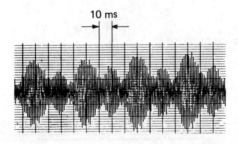

Fig. 10.60   Structure-borne sound signal of magnetic noise (10.38).

166

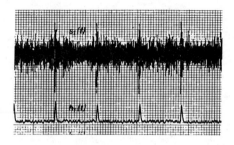

*Fig. 10.61   Airborne sound signal $s_1(t)$ of grinding noises and processed signal $h_1(t)$
(10.38).*

synchronised to the revolutions is typical for grinding noise. They are produced by striking a spring-mounted vibrating body (signal variation $s_1(t)$ in Fig. 10.61).

The signal processing of grinding noises is shown in Fig. 10.62. The grinding noise components are let through a high-pass filter. Frequencies under 6 kHz are suppressed. A rectangular signal ($h_1(t)$ in Fig. 10.61) corresponding to the signal peaks can be derived from the envelope by means of a threshold comparator. Its existence is used for the evaluation as well as its revolution synchronicity in order to distinguish it from other bearing noises which are also periodic but not synchronous with the revolutions.

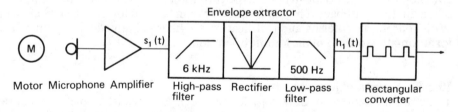

Fig. 10.62   Signal processing of grinding noise (10.38).

(c) Bearing noises
With bearing noises similar high peaks occur as with grinding noises. However these are not synchronous with the revolutions and usually occur repeatedly during one motor revolution. The processing is similar to that of grinding noises, the lack of revolution synchronicity being used as a distinguishing feature. Fig. 10.63 shows the overall testing system for all three fault categories.

10.2.4   Monitoring of forging hammers:
The red-hot workpiece treated by a forging hammer is completely formed when the upper and lower die meet each other for the first time. This die-to-die blow can be clearly distinguished from normal impact noises and is taken by the operator as a signal to end the operation. The noise level puts a considerable strain on the operator and is injurious to his health. A brief description will now be given of an acoustic signal processing solution to this problem. An analysis of the air-borne dissipated to the surroundings by the forging hammer which is excited to vibrations by the rapidly changing forces produces the spectrum shown in Fig. 10.64 (10.39).

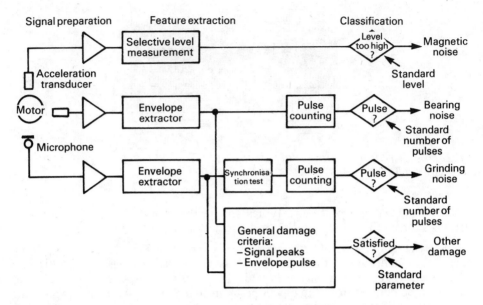

Fig. 10.63   Acoustic testing system for an induction electric motor (10.27).

The forging hammer is excited to vibrations of a few kHz. The time for the sound emissions is characteristically of the order of 100 ms (10.40). The larger amplitude of the die-to-die impact at higher frequencies can be used for the signal processing. With appropriate filtering and rectification a signal is obtained which can be used for automatic stopping of the forging operation.

10.2.5   Acoustic object recognition

There are difficulties in knowing whether to classify acoustic object recognition systems as optical or acoustic systems. Since in the recognition system described

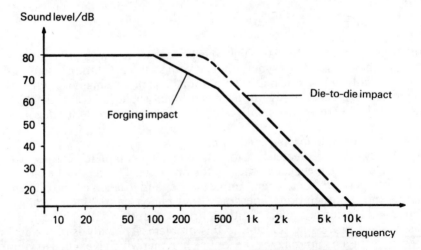

Fig. 10.64   Sound spectrum of forging impact and die-to-die impact.

168

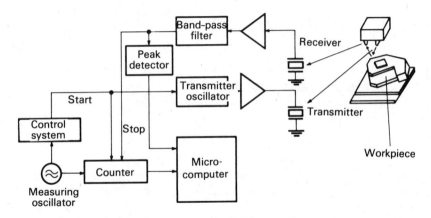

Fig. 10.65   System for acoustic object recognition (10.41).

below, use is made of acoustic transducers, it will be placed under the heading of acoustic systems.

The principle of the recognition system is shown in Fig. 10.65. For object recognition the sound emitted by an ultrasonic transmitter (e.g. a quartz oscillator) is reflected at the object and received by a microphone. A distance picture of the object can be constructed from the transit time between the transmitted and received sound. The radiation and reception angle of the transmitter and receiver is less than 10 degrees so that a good resolution is obtained. It is possible to cover the area of the object by making scanning movements of the transmitter and receiver.

The transit times between the transmitted and received signal are determined in a counter which is started at the beginning of the transmitted signal and stopped by the signal reflected at the workpiece. A microcomputer reconstructs the distance image of the object (Fig. 10.66) from the transit time, the received signal value and the associated scanning position of the transmitter and receiver. Similar pictures are obtained to those of the light section method in the optical field.

One difficulty with this system is encountered with very complex objects with complicated surface structures. Reflection phenomena can lead to ambiguities here.

Fig. 10.66   Acoustically generated distance image (10.41).

All known methods of image processing can be used for further processing of the acoustically generated distance image.

10.3 Tactile sensor systems

As explained in Chapter 6, there are two main application areas for tactile sensors:

1. Recognition tasks
2. Control of movement operations (joining, path following)

The dominace of optical systems for recognition tasks predisposes tactile sensors for the control of movement operations. In the industrial sector the necessary feeling and touching for this can be described by a succession of path-related and force-related operations.

It is especially clear with the following examples that the actual measuring technology is limited to known methods and principles of measuring transducers. The difficulties in solving tactile sensor tasks lie in finding processing algorithms and suitable special design solutions for the transducers.

10.3.1 Gripping systems for workpiece recognition

Workpiece contours can be determined by specially designed grippers and the result used for recognising the workpiece. In the simplest case it is a matter of moving pins, as shown in Fig. 10.67, which reproduce the workpiece contour (10.42). In more complicated cases the gripper can even be modelled on the human hand with moving thumb and fingers (10.43).

The positions x_i of the pins (Fig. 10.67) are the components of the feature vector $c = (x_1, x_2, \ldots, x_n)$. As with optical sensors the workpiece can be gripped in the programming phase in order to learn the geometrical features. In the actual working phase a workpiece is recognised by comparing the measured features with the learned features, e.g. by means of one of the classification algorithms mentioned in Chapter 8.3. The components of x_i of the feature vector can be determined by any of the transducers for force or path measurement listed in Section 6.3.

A simple implementation of a tactile gripping system with elastometers (Section 6.3) is shown in Fig. 10.68 (10.44).

A matrix field of 8×10 pins scans the workpiece. These moving pins alter the local pressure on a piece of plastic whose resistance drops with increasing pressure: this means a larger current flow at the contact points with the workpiece. A matrix impression of the workpiece can be taken from the fixed pins. The LEDs shown in Fig. 10.68 serve as control indicators for the matrix representation.

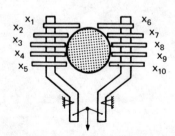

Fig. 10.67   Gripper with shaping jaws.

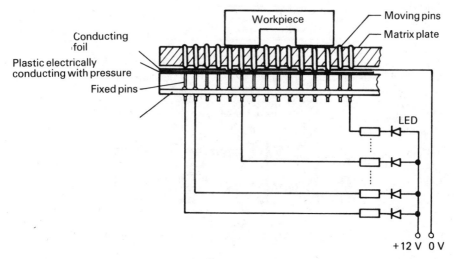

Fig. 10.68   Tactile transducer with elastomers (10.44).

10.3.2   Weight-sensitive gripping system

With many workpieces being handled the gripping force has often to be adapted to them. For instance, when gripping fragile parts the gripping force must be more carefully regulated than when gripping heavy castings.

The principle of a weight-sensitive gripping system will now be presented (10.45). This does not only enable the workpiece to be gripped, picked up and set down as a function of its weight but at the same time makes it possible to recognise the workpiece by determining its weight.

The principle of the weight-sensitive gripping system is shown in Fig. 10.69. The workpiece is gripped by both jaws and lifted up. At the same time a force gauge measures the contact pressure of the jaws and a weight tranducer determines the weight. Both transducers consist of a spring arrangement with potentiometer pick-offs. At the beginning of gripping the force is so small that a slight sliding motion

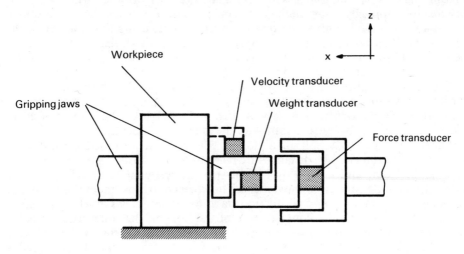

Fig. 10.69   Principle of the weight-sensitive gripping system (10.45).

171

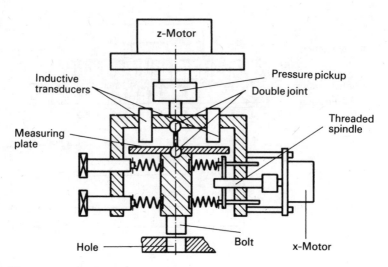

Fig. 10.70 Principle of a joining system (10.46).

between the jaws and the workpiece is premitted. The relative movement between workpiece and gripper is determined by means of a velocity tranducer at the gripping system mechanically or magnetically coupled to the workpiece. Here is is sufficient to have the binary information whether the workpiece is slinding or not. If the relative velocity is zero the unknown coefficient of friction between work-piece and jaws is determined in a division circuit from the quotient of the weight and contact force. This value is multiplied by the weight and provides the optimum force i.e. the minimum force necessary for gripping.

If the workpieces have different weights the weight value determined during the gripping operation can be used at the same time for recognition.

10.3.3   Sensors for joining (mating parts)

Mating parts is a common task during assembly. For instance when a bolt is inserted into a hole by a human the following procedure is usual: the bolt is guided to the hole by visual control, set directly over the hole using the mobility of the fingers and then using the sense of touch inserted in the hole. An unpractised person sometimes needs more than four seconds for inserting a bolt with a diameter of 20 mm and a clearance of 0.013 mm (10.46).

According to Section 6.2 the inclination of the bolt, as well as the force necessary for joining, are important variables during a joining operation. A transducer for measuring these variables is shown in Fig. 10.70 (10.46). This can be used to illustrate the basic procedures adopted for the measuring process.

The inclination of a bolt in a holding device is determined by inductive proximity transducers. The distances are measured to a measuring plate suspended at a double joint and guided by helical springs. A signal is derived from the inclination for the correction movement. This correction movement is executed by stepping motors in the x- and y-direction via threaded spindles. Only the x-motor is shown in Fig. 10.70. The contact force of the bolt on the hole (axial force) is determined by a pressure pick-up using a wire strain gauge principle. This signal controls the movement of the motor in the z-direction. If the contact force exceeds a preset threshold (bolt is inserted in the hole, bolt is crooked) the z-motor is stopped.

172

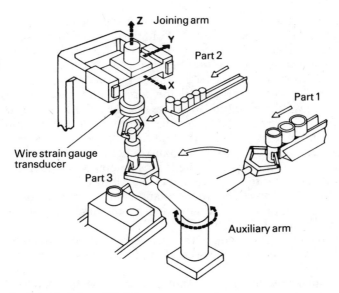

Fig. 10.71   Automatic joiner (10.47).

This is only one possible example of a transducer for a joining system. A more complex sensor system will now be presented for the same task (10.47) Fi.g 10.71 shows the overall view of the system.

An auxiliary arm takes the first part from a delivery system, brings it to the joining arm and holds the part there in the assembly position. The joining arm grips the bolt (part 2) and inserts it in part 1. After this assembly operation has been completed the joining arm takes both parts and if necessary inserts them in a third part.

The gripper of the joining arm is coupled in a flexible manner to the arm by means of a special wire strain gauge transducer (Fig. 10.72). Four wire strain gauges on crossed leaf springs enable the force components to be measured in the three

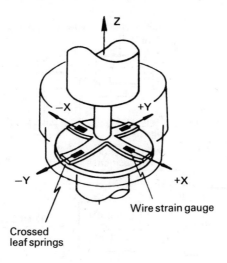

Fig. 10.72   Wire strain gauge transducer of the joining arm.

173

axial directions x, y and z. These measurement signals are used to carry out the three dimensional joining movement without the help of a computer by means of a sequence control system.

The joining process and the accompanying measuring signals will now be examined in more detail. At the beginning of the process part 1 is moved in the z-direction by the joining arm until the force component monitored by the transducer exceeds a threshold and indicates that the bolt has been placed over part 2. In the subsequent search phase the bolt is moved in the y-direction. Since the maximum permissible deviation of the bolt from the hole is ± 2 mm it is ensured that the hole will be found in only one axial direction with this movement. If the bolt partly overshoots the hole it is tilted. This generates a characteristic change in the transducer signal for the z-component. The tilting of the bolt is measured and from this a correction movement in the x- and y-direction derived. This aligns the bolt with its axis parallel to the hole and by moving the joining arm in the z-direction it is inserted into the hole.

The time taken to insert a bolt with a diameter of 20 mm into a hole with 0.03 mm clearance is between 1 and 3 seconds and so on average is shorter than that taken by humans.

The automatic joiner does not need any computer. All the search operations and movement sequences are controlled according to the sensor signals by a sequence control system programmed into a ROM.

10.3.4 Deburring system with tactile control

In manufacturing technology deburring is understood as the removal or reduction of excess material which occurs in the form of opverlaps at edges of workpieces. The commonest task is the deburring of castings. Typical workpieces are gearboxes, engine blocks or sewing machine cases. With small-scale and medium-scale production this work is carried out manually with the help of rotating grinding tools. This bodily strenuous activity has to be performed under difficult conditions (noise, dust, heat). Only humans have the necessary visual control and the sense of touch in order to recognise even gross deviation in shape and to correct accordingly. Programmable manipulation systems (industrial robots) are in principle also well equipped to handle these operations. Because of casting

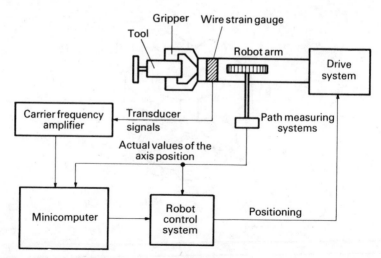

Fig. 10.73   Design of the deburring system (10.49).

174

tolerances, casting errors and machining tolerances deviations can occur at the edges and surfaces of workpieces of the same type of up to 20 mm. This means that the deburring operation cannot be carried out along a previoulsy defined programmed path. The forces acting on the workpiece must, therefore, be measured by tactile sensors in order to control the manipulation system by means of an appropriate movement algorithm.

The tactile transducer consisting of force and moment gauges is usually situated directly behind the grinding machine on the arm of the manipulation system where it measures the entire force flux. The inertia of the transducer and the parts in front of it leads to an undesirable force measurement on acceleration. These acceleartion forces must, therefore, be kept small in comparison with the commonly occurring cutting forces. The transducer has to measure three forces and moments perpendicular to each other separated according to individual components. Because of the frequently occurring chatter vibrations during deburring the transducer has to be very rigid. Piezoelectric or wire strain gauge transducers are most suitable for this. Wire strain gauges have the advantage of

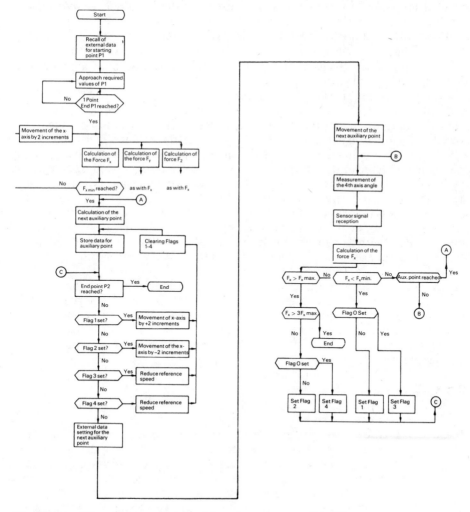

Fig. 10.74   Flow diagram for the deburring program (10.49).

simple temperature compensation and low price. The subsequent electronic processing takes place with the help of carrier frequency amplifiers (10.48).

A sensor-controlled deburring system will now be presented which is equipped with a wire strain gauge for measuring cutting forces and torques (10.49).

Such wire strain gauges attached to springs of the most varied types for measuring forces and moments are standard transducers in measuring technology (10.50, 10.41). By combining several transducers for single measuring components a multi-component transducer is obtained (10.49, 10.50).

The principle of the deburring system is shown in Fig. 10.73. The signals from the wire strain gauges arranged in a half bridge on the measuring devices are input to a minicomputer via a carrier frequency amplifier. Apart from these force and moment signals the computer also receives information on the position of the axes from the path measuring system of the industrial robot. The required path and speed values for the drive of the manipulation system are calculated by the computer from these signals.

When the system is programmed the coordinates of the required path to be travelled are stored in the computer. This can either be done analytically by presetting data or in the teach-in procedure by following the required curve. In addition the maximum and minimum permissible force components F_{max} and F_{min} are input. The presetting data is input as part of a VDU interactive dialogue.

The program flow diagram of Fig. 10.74 shows the procedure for automatic following of a straight path.

Beginning at the learnt starting point P_1 the axial co-ordinates of the manipulation system are adjusted until the preset minimum force has been reached. The rate of cut or rate of feed is changed when the maximum force F_{max} is exceeded as a function of the flags set in the program (flags 1-4 in Fig. 10.74).

If a force is more than F_{max} but smaller than $3F_{max}$ then when the next auxiliary point is encountered either the reference speed is reduced or depending on whether flag O is set (Fig. 10.74) the required position value of the relevant axis is reduced by two increments. Similarly if a force is less than F_{min} the required speed for the next auxiliary point is increased or the reference value of the axial position increased by two increments. If the flag O is set the manipulation system reacts to the threshold being exceeded with a change in the feed rate and if the flag O is not set with a pre-programmed change in path. The operation terminates when the end point P_2 is reached.

This system is still at the stage of undergoing trial operations with the relatively simple task of following a straight edge. However, by further development of the computer program and improvements in the force and moment transducers in the sense of a better separation between the individual force and moment components it will also be possible in the future to handle more complicated workpiece shapes in industrial applications.

10.3.5 Tactile assembly system

In Section 10.1.4 an assembly system is described which makes it possible to carry out the complex operations involved in assembling vacuum cleaner parts under optical and tactile sensor control. In this case there are no attempts made to simplify the process. For instance, the parts to be assembled lie disordered on a stack. The hardware necessary is, therefore, extensive, elaborate and costly. This is why this system is not yet suitable for industrial application. However, it shows what is in principle possible today and the difficulties of automatic assembly.

An assembly system will now be presented which with a single tactile sensor of simple design also carries out complex assembly tasks and is already used in many

Fig. 10.75   Overall view of the assembly system (Source: IPA Stuttgart).

practical applications (10.52, 10.53). In contrast to the system presented in Section
10.1.4 a defined position of the parts is obtained by means of simple design measures
and devices and in this way the cost considerably reduced.

The assembly system is characterised by two assembly arms arranged vertically
in the z-direction which can be moved by motors in all three axial directions with
an accuracy of 0.05 mm. The parts to be assembled are first ordered and then
presented to the grippers of the assembly arms for insertion. Fig. 10.76 shows the
construction of the transducer in the assembly arm. The gripper joint is constructed
in such a way that no deflections are possible in any of the three axial directions.
These deflections are measured by three path transducers. Since a deflection
actuates springs whose reaction force is proportional to the travel of the spring the
measured paths can be used to determine the reaction forces. As shown in Fig. 10.76
the gripper can be shifted over a small area of the x-y plane on a slide plate so that it
can move in the x-y plane. This means that apart from the measurement of the
reaction forces and position deviations necessary for the control system a self-
centering system is obtained.

By means of good prepositioning of the parts to be assembled it is sufficient for
the joining operation to evaluate the signal of the path transducer in the z-direction
(parallel to the joining direction). In this way it is unnecessary to simultaneously
process several measurement signals.

The assembly system is programmed for the different operations by means of a
control stick. This makes it possible to simulate various movement operations

177

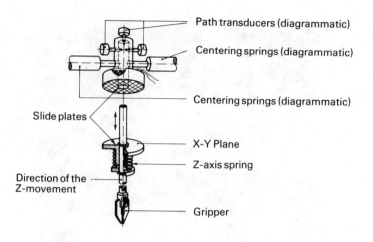

Path transducers (diagrammatic)

Centering springs (diagrammatic)

Centering springs (diagrammatic)

Slide plates

X-Y Plane

Z-axis spring

Direction of the
Z-movement

Gripper

Fig. 10.76   Assembly arm transistor (10.52).

needed later for the task. Alternatively, the movement sequence can be numerically input in the form of coordinate values.

In the actual assembly operation the value measured by the path measuring system is compared with the programmed reference value. Depending on the result of the comparison the assembly is either continued based on the previously programmed instructions or terminated. When assembling parts for instance, undrilled parts are recognised as a result of the force measured while inserting and the operation broken off. A minicomputer with a memory of 8k words (16 bits) is sufficient for controlling the assembly operations.

Some idea of the performance of the system can be obtained from the following brief descriptions of industrial applications.

When assembling the 48 type shanks of a typewriter the delivery rate of the manipulation system is double that obtained by a human performing the same task (10.52). In another application the assembly of a typewriter ribbon cassette consisting of the ribbon, two spools, two attachment parts and two casing covers only takes nine seconds. This makes the assembly about six times faster than a human. The assembly system brings benefits when it handles a few parts with frequently reoccurring operations.

This example shows how complex assembly tasks can be carried out fully automatically with relatively low-level sensor and electronic hardware by means of suitably programmed operation sequences in conjunction with simple mechanical aids for isolating, ordering and rough positioning.

178

11. Outlook

Forecasts of technological developments often either make the mistake of over-estimating the possibilities of the near future of underestimating the possibilities of the distant future. What is certain, however, is that intelligent measuring systems will find increasing application in the automation of industrial processes.

In the field of optical sensors the first industrial applications for the relatively simple case of binary picture processing have made their mark. However, it should not be overlooked that also here many systems are still in the stage of laboratory development.

With very many industrial applications, it is not possible right from the beginning to generate binary pictures. Therefore, in the future grey-tone picture processing will be of increasing importance. Examples where grey-tone picture processing is necessary are the recognition of an object aainst a dirty background (e.g. picking up workpieces from pallets) and following a weld.

Even when implementing procedures for grey-tone picture evaluation it will be necessary to tread the familiar path of:

- dividing the task into practicable individual intermediate steps,
- reducing the amount of information (e.g. by suitable illumination procedures such as light section methods, etc.).
- simplifying existing processing algorithms of pattern recognition or the development of new methods suitable for industrial application in terms of memory requirements and processing speed.

It is found that the optical sensor tasks in the industrial sector can be classified into a few task categories. These go right across the board of industrial activity:

- Position determination and relative positioning,
- Object recognition (classification),
- Visual inspection.

Because the solution is not specific to any branch of industry, the tendency in the future will be to develop modular systems in the optical field. Using a suitable combination of software and hardware, the system will have built-in flexibility and versatility. Here one is mainly talking of software modifications which do not involve changing the overall system.

It can be expected that grey-tone picture processing will reach a similar stage of development to binary picture processing (laboratory testing, first industrial applications) in the mid eighties.

By this time tactile sensors will also have found increasing application in industry. The focus of research here lies in the development of suitable algorithms for the transducer signals. Key tasks here are the joining of parts during assembly and the following of edges (e.g. deburring).

The main tasks in the future of acoustic sensors are early warning of faults and machinery surveillance as well as acoustic quality control in production.

179

Procedures are being developed in these areas at the present time. In a few years time they will be increasingly used for simpler tasks (e.g. small engines). The first applications to more complex objects (e.g. internal combustion engines) can also be expected from the mid eighties.

The key role of sensor technology and its rapid spread in industry is best illustrated by the result of a survey of experts conducted in the USA (11.1).

According to this in 1982 about 25% of assembly tasks will be monitored by sensors. Industrial robots will then be used in about 5% of the tasks in this area. This percentage will have risen to about 15% by 1987.

In 1985 about 7% of assembly systems will have sensor monitoring. According to this forecast around the year 1990 sensor technology will be so advanced that automatic assembly systems will come near to emulating the performance of humans.

12. Literature

The literature references are ordered according to chapter and within a chapter numbered in order of occurrence.

Chapter 1

(1.1)   Hart, H.   Introduction to measuring technology. VEB-Verlag Technik, Berlin 1978.

Chapter 2

(2.1)   BMFT   Main objectives of the BMFT in promoting the the research programme "Humanisation of working life". BMFT communications 11/1976, p. 103/104.

(2.2)   Lang, H.   Programmable workpiece recognition device. Assembly and Handling technology (1976), Vol. 3, p. 95-57.

(2.3)   Rein, H.   Introduction to human physiology. Springer-Verlag Berlin, 1941.

Chapter 4

(4.1)   Grosch, Al.   Camera tubes with a silicon target as the image sensor. Elektronik (1974), Vol. 2, p. 55-58.

(4.2)   Limann, O.   Television technology without ballast. Verlag Munich, 1973.

(4.3)   König, S.   Principle of operation and application of charge coupled devices. Fernseh- and Kino-technik, Vol. 29 (1975), part 4 p. 107-110.

(4.4)   Pfleiderer, H. J.   Optoelectronic sensors with CCDs. Elektonik (1975), Vol. 4, p. 88-92.

(4.5)   Theile, R.   Television technology. Vol. 1 Basic Principles, Springer Verlag, Berlin, Heidelberg, New York, 1973.

(4.6)   Holeman, B. R.   Infrared and visible radiation detectors for imaging and non-imaging applications. AGARD lecture series on opto-electronics (Sept. 1974).

(4.7)   Cope, A. D.   The television camera as a system component.
        Gray, S.   In: Photoelectronic imaging devices Volume 2.
        Hutter, E. C.   Plenum Press New York, London, 1971.

(4.8)   Data Sheet Reticon Corp. Photosensor area array RA-100×100.

(4.9)	Stupp, E. H. Levitt, R. S.	The Pumbicon. In: Photoelectronic imaging devices. Volume 2, Plenum Press, New York, London 1972.
(4.10)	Haendle, J. Horbaschek, H. Alexandrescu, M.	The high-resolution X-ray television and the high-resolution video memory. Electromedica (1977). Vol. 3-4, p. 87.
(4.11)	Hall, A. J.	Evaluation of signal-generating image tubes. In: Photoelectronic imaging devices. Volume 2, Plenum Press, New York, London, 1972.
(4.12)	Bretschi, J. König, M. Schief, A.	A reduction of information in optical sensors of industrial robots. 2nd CISM-IFTOMM symposium on theory and practice of robots and manipulators. Warsaw, September 1976.
(4.13)	Foith, J. P.	Review of the technology and application of digital image sensors – Internal Report IITB Karlsruhe.
(4.14)	Foith, J. P.	Image processing for orientation and position determination in applications of industrial robots. IITB Communications 1977, p. 32-40.
(4.15)	Shirai, Y.	Recognition of polyhedrons with a range finder. Pattern recognition 4 (1972), p. 243-250.
(4.16)	Röcker, F.	On the problem of automatic analysis of three-dimensional scenes. Report No. 33, Research Institute for information processing and pattern recognition (FIM), Karlsruhe, Jan. 1976.
(4.17)	Bretschi, J. Gudat, H. Lübbert, U. Paul, D. Schief, A.	Industrial sensors for the automation of the human sense faculties in production. FhG Reports 1/77, p. 4-12, Fraunhofer-Gesellschaft, Munich, 1977.
(4.18)	Bretschi, J.	Sensors for automating production. wt-Z. ind. Fertig. 67 (1977), p. 393-396.
(4.19)	Bretschi, J. König, M.	High-performance sensors for industrial robots – objectives , systems, technical consequences. Conveying and lifting (Specialist section assembly and handling technology) 26 (1976), Volume 13, p. 2-4.
(4.20)	Peipmann, R.	Principles of industrial recognition. VEB Verlag Technik, Berlin, 1975.

Chapter 5

(5.1)	Gudat, H.	Possibilities of acoustic quality control in industrial production. Specialist reports, Measurement, control, regulation, Volume 1, Interkama Congress 1977.
(5.2)	Sessler, G. M. West, I. E.	Foil electret microphones. Journ. Acoust. Soc. of America 40 (1966), p. 1433.
(5.3)	Reichardt, W.	Principles of electroacoustics. Akadem. Verlagsgesellschaft Geest & Portig, Leipzig, 1960.

(5.4)	Günther, B. C. Hansen, E.-H. Veit, I.	Industrial Acoustics. Lexika-Verlag 1978.
(5.5)	Veit, I.	Industrial acoustics in a nutshell. Vogel-Verlag 1974.
(5.6)	Heckl, M. Müller, H. A.	Handbook of industrial acoustics. Springer-Verlag, Berlin 1975.
(5.7)	Veit, I.	Industrial measurement in acoustics and vibration technology – a review. Technisches Messen atm (1977), Volume 5, p. 163-173.
(5.8)	Schulz, D.	Linear and non-linear processing of amplitude-modulated stochastic processes for acoustic quality control and early warning of faults. Thesis Technical University of Karlsruhe, 1976.
(5.9)	Schulz, D.	Acoustic testing of gears. IITB Communications 1976, Karlsruhe, p. 2-5.
(5.10)	King A. I.	The measurement and suppression of noise. Chapman & Hall, London 1965, p. 100-107.
(5.11)		DIN 1320.
(5.12)	Mitchell, L. D. Lynch, G. A.	Origins of noise. Machine design. May 1, 1969.
(5.13)	Pollehn, W.	Acoustic methods for controlling blowing operations in steel converters. Stahl und Eisen 88 (1968), No. 9, p. 457-459.

Chapter 6

(6.1)	Schweizer, M.	Sensors for program-controlled handling systems. f + h – fördern and heben 27 (1977), No. 4, Specialist section mht, p. 22-27.
(6.2)	Nervin, J. L. Whitney, D. E.	Computer-controlled assembly. Scientific American, Feb. 1978, p. 62-74.
(6.3)	Drake, S. H. Watson, P. C. Simunovic, S. N.	High-speed robot assembly of precision parts using compliance instead of sensory feedback. Proc. 7th Int. Symp. on Industrial Robots, Tokyo, Japan, Oct. 1977.
(6.4)	Rohrbach, Ch.	Handbook for the electrical measurement of mechanical variables. VDI-Verlag, Dusseldorf 1967.
(6.5)	Haug, A.	Electronic measurement of mechanical variables. Carl Hanser Verlag, Munich 1969.
(6.6)	Trade Publication	Dynacon Industries Inc., Leonia, New Jersey.
(6.7)	Kuist, Ch. H.	Those flexible, formable current carrying plastics, Automation, Oct. 1976, p. 54-57.
(6.8)	Nabereit, H.	Polymer force transducers. Communication from the Institute for the principles of electrotechnology and electrical measuring technology at the Technical University of Brunswick, INTERKAMA 1977.

Chapter 7

(7.1)   Düll, E. H.   The microprocessor in control technology. Regelungstechnische Praxis 1974, Vol. 11, p. 279-285.

(7.2)   Osborne, A.   Introduction to microcomputer technology, te-wi Verlag GmbH, Munich 1977.

(7.3)   Soucek, B.   Microprocessors and microcomputers. John Wiley & Sons, New York, London, Sidney, Tokyo, 1976.

(7.4)   Korn, G. A.   Microprocessors and small digital computer systems for engineers and scientists, McGraw-Hill Book Company, 1977.

(7.5)   Hilberg, W. / Piloty, R.   Microprocessors and their applications. R. Oldenbourg Verlag, Munich, Vienna, 1977.

(7.6)   Blomeyer-Bartenstein, H. P.   Microprocessors and microcomputers. Siemens AG, Munich, 1977.

(7.7)   Kobitzsch, W.   Microprocessor design and operation Part I. Basic principles. Part II Applications and examples. R. Oldenbourg Verlag, Munich, 1978.

(7.8)   Leventhal, L. A.   Introduction to microprocessors: Software, Hardware, Programming. Prentice Hall, INC., Englewood Cliffs, New Jersey, 1978.

(7.9)   Gräf, R. / Kammerer, J.   Electronics IV C Microcomputer design, applications, programming. Richard Pflaum Verlag KG Munich, 1979.

(7.10)   Soeiser, A. P.   Digital computer systems. Springer-Verlag, Berlin/Gottingen/Heidelberg, 1961.

(7.11)   Münchrat, R.   Trends in minicomputer development. Elektronik 1073, Vol. 10, p. 345-352.

Chapter 8

(8.1)   Niemann, H.   Methods in pattern recognition. Akademische Verlagsgesellschaft, Frankfurt 1974.

(8.2)   Meyer-Brötz G. / Schürmann, J.   Methods of automatic character recognition. R. Oldenbourg Verlag, Munich–Vienna, 1970.

(8.3)   Tou, J. T. / Gonzales, R. C.   Pattern recognition principles. Addison-Wesley Publishing Company London, Amsterdam, Tokyo, 1974.

(8.4)   Schulz, D.   Linear and non-linear processing of amplitude-modulated stochastic processes for acoustic quality control and early warning of faults. Thesis Technical University of Karlsruhe, 1976.

(8.5)   Winkler, G.   Stochastic systems, analysis and synthesis. Akademische Verlagsgesellschaft Wiesbaden, 1977.

(8.6)   Lange, F. H.   Signals and systems. Volume 3, Random processes, VEB Verlag Technik Berlin 1971.

(8.7)   Veit, I.   Industrial acoustics. Vogel Verlag Wurzburg, 1974.

(8.8)	Bretschi, J.	Linearisation of measurement transducers illustrated by the example of wire strain gauges. Technisches Messen atm 1976, Vol. 11, p. 349-356.
(8.9)	Schleifer, W. D.	Diode transducers for empirical functions using operational amplifiers. Internationale Elektronische Rundschau, 1967, No. 11, p. 279-280.
(8.10)	Schleifer, W. D.	Diode transducers for empirical functions. Internationale Elektronische Rundschau 1967, No. 9, p. 235-238.
(8.11)	Best, R.	The generation of nonlinear functions with analog function generators. Der Elektroniker 5 (1970), p. 247-254.
(8.12)	Pavlidis, T. Fang, G. S.	Application of pattern recognition to fault diagnosis of internal combustion engines. Tech. Report No. 98, Sept. 1971, Computer Science Laboratory, Princeton University.
(8.13)	Barschdorff, D. Hensle, W. Stühlen, B.	Noise analysis for early warning of faults on stationary turbomachinery as a problem in pattern recognition. Technisches Messen atm Vol. 5, (1977), p. 181-188.
(8.14)	Peterson, R. H. Ackermann, A. D. Zelensiki, R. E.	Acoustic signal analysis for noise source identification in mechanisms. IBM J. Res. Develop., Vol. 6 (1972), p. 249-257.
(8.15)	Hoffman, R. L. Fukunaga, K.	Pattern recognition signal processing for mechanical diagnostic signature analyses, IEEE Trans. Comput., Vol. C-20 (1971), p. 1095-1100.
(8.16)	Schulz, D.	Acoustic testing of gears. IITB Communications 1976, Karlsruhe, p. 2-5.
(8.17)	Brigham, E. D.	The fast Fourier transform. Prentice-Hall Inc., Englewood Cliffs, New Jersey, 1974.
(8.18)	Bergland, G. D.	A guided tour of the fast Fourier transform. IEEE Spectrum, July 1969, p. 41.
(8.19)	Steinbrenner, H. Müller, E. Lenartz, H.	Automatic order analysis of engine vibrations, Elektronik 1967, Vol. 11, p. 343-349.
(8.20)	Schlitt, H. Dittrich, F.	Statistical methods of control technology. BI Hochschultaschenbücher Bibliographisches Institut, Mannheim, Vienna, Zurich, 1972.
(8.21)	Bendat, J. S. Piersol, A. G.	Measurement and analysis of random data. John Wiley & Sons, New York, 1966.
(8.22)	Oppenheim, A. V. et al.	Nonlinear filtering of multiplied and convolved signals. Proc. IEEE Vol. 56, No. 8, Aug. 1968, p. 1264-1291.
(8.23)	Oppenheim, A. V. Schafer, R. W.	Digital signal processing. (Chapter 10), Prentice-Hall, New Jersey, 1975.
(8.24)	Günther, B. C. Hansen, K.-H. Veit, I.	Industrial acoustics. Lexika-Verlag, Grafenau, 1978.

(8.25)	Zaveri, K. Phil, M.	Acoustic examination of a piston drill. Bruel & Kjaer Tech. Rev. No. 3 (1974).
(8.26)	Thomas, D. W. Wilkins, B. R.	The analysis of vehicle sounds for recognition. Pattern recognition, Pergamon Press Vol. 4 (1972), p. 379-389.
(8.27)	Foith, J. P.	Position recognition of randomly orientated workpieces from the shape of their silhouettes. Proc. 8th Int. Symp. on Industrial Robots, West Germany, 1978, p. 584-599.
(8.28)	Author's collective	Half-yearly report I/1976 of the handling systems working group chapter 23.2 BMFT Project 01 VC 044 – B 13 TAP 002.
(8.29)		Engineer's workshop handbook. Theoretical principles, 28th edition, Verlag von Wilhelm Ernst & Sohn, Berlin 1955, p. 671 et seq.
(8.30)	Hasegawa, K. Masuda, R.	On visual signal processing for industrial robots. Proc. 7th Int. Symp. on Industrial Robots Tokyo, 1977, p. 543-550.
(8.31)	Duda, R. O. Hart, P. E.	Pattern classification and scene analysis. John Wiley & Sons, New York, Toronto, London 1973.
(8.32)	Meyer-Eppler, W. Darius, G.	The autocorrelation of flat two-dimensional pattern pictures. NTF, Vol. 3, 1956, p. 40-46.
(8.33)	König, M.	Contactless recognition and position determination of objects by incoherent optical correlation. Krausskopf-Verlag, Mainz, 1977.
(8.34)	Yoda, H. et al.	Direction coding method and its application to scene analysis. Proc. 4th Int. Joint Conf. on Artificial Intelligence, 1975, p. 620-627.
(8.35)	Agrawala, A. K. Kulkarni, A. V.	A sequential approach to the extraction of shape features. Computer Graphics and Image Processing 6 (1977), p. 538-557.
(8.36)	Kulkarni, A. V.	Sequential shape feature extraction from line drawings. Proc. Pattern Recognition and Image Processing 1978, p. 230-237.
(8.37)	Peipmann, R.	Principles of industrial recognition. VEB Verlag Technik, Berlin, 1975.
(8.38)	Batchelor, B. G.	Pattern recognition ideas in practice. Plenum Press, New York, London, 1978.
(8.39)	Kulikowski, C. A.	Pattern recognition approach to medical diagnosis. IEEE Trans, on Syst. Scie. and Cyb. 3 (1970), p. 173-178.
(8.40)	Foith, J. P. König, M.	Black and white image sensors in production technology – a review. Technisches Messen atm, Vol. 4, (1978), p. 135-139.
(8.41)	Rosen, C. et al.	Exploratory research in advanced automation. Stanford Research Inst., 5th Rep., 1976.

Chapter 9

(9.1)	No author	Trade publication Pfaff Pietzsch Industrial Robots GmbH, Ettlingen.

| (9.2) | Auer, B. A. et al. | Industrial robots and their practical application. Lexika-Verlag, Grafenau, 1979. |
| (9.3) | Geisselmann, H. | The television sensor and its link-up with an industrial robot. 8th Int. Symp. on Industrial robots, Stuttgart 1978, p. 175-177. |

Chapter 10

(10.1)	Warnecke, H. J.	Reflections on the automation of handling. 3rd Int. Symp. on Industrial Robots, March 1973, Zurich, Verlag Moderne Industrie, Munich.
(10.2)	Giesen, K.	Conference on workpiece handling in automatic production, March 1970, Haus der Technik, Vulkan-Verlag Essen.
(10.3)	Brödner, P. Schacks, P.	Development and application of industrial robots within the framework of the West German Government's research programme "Humanisation of working life". 8th Int. Symp. on Industrial Robots, 1978, Stuttgart, Int. Fluidics Services, Kempston, England.
(10.4)	Schraft, R. D.	The design of industrial robots – a survey of the international market. 3rd Int. Symp. on industrial robots, March 1973, Zurich; Verlag Moderne Industrie, Munich.
(10.5)	Author's collective	Study of technical aids for the working process. March 1974, BMFT Research Report T 74-03. ZLDI, Munich.
10.6)	König, M.	Contactless recognition and position determination of objects by incoherent optical correlation. Krausskopf-Verlag BmbH, Mainz, 1977.
(10.7)	Author's collective	New handling systems as technical aids for the working process. Annual Report 1974, Item 19 of the handling systems working group. BMFT project 01 VC 044 – B 13 TAP 002.
(10.8)	Author's collective	New handling systems as technical aids for the working process. Annual report 1976/I, Item 23.2 of the handling systems working group. BMFT project 01 VC 044 – B 13 TAP 002.
(10.9)	Author's collective	New handling systems as technical aids for the working process. Annual report 1977/II, Item 9.7 of the handling systems working group. BMFT project 01 VC 044 – B 13 TAP 002.
(10.10)	Stratemeier, U.	Optoelectronic orientation recognition of workpieces with small feature differences. Proc. 8th International Symposium on Industrial Robots 1978, Vol. 2, p. 734-743, Int. Fluidics Services, Kempston, England.
(10.11)	Lang, H.	Programmable workpiece recognition device. Montage und handhabungstechnik 3 (1976), p. 95-97.

(10.12)	Bretschi, J.	German patent application p. 25 13 655.4.
(10.13)	Bretschi, J.	A microprocessor-controlled visual sensor for industrial robots. The industrial robot 3 (1976), Vol. 4, p. 167-172.
(10.14)	Author's collective	New handling systems as industrial aids for the working process. Annual report 1977/II, Item 9 of the handling systems working group. BMFT project 01 VC 044 − B 13 TAP 002.
(10.15)	Geisselman, H.	Television sensor for workpiece recognition, position measurement and quality control. IITB communications 1977, p. 27-31.
(10.16)	Geisselmann, H.	The television sensor and its link-up with an industrial robot. Proceed. 8th Int. Symp. on industrial robots, 1978, Vol. 1, p. 165-180. Int. Fluidics Services, Kempston, England.
(10.17)	Heginbotham, W. B. et al.	The Nottingham "Sirch" assembly robot. First Conference on industrial robot technology. University of Nottingham, 1973.
(10.18)	Karg, R.	A flexible opto-electronic sensor. Proc. 8th Internat. Symposium on industrial robots 1978, Vol. 1, p. 218-229. Int. Fluidics Services, Kempston, England.
(10.19)	Lanz, D. E.	Sensor for orientation and shape recognition. Specialist reports measuring, control, regulation, INTERKAMA Congress 1977, Vol. 1, p. 95-106. Springer Verlag, Berlin, Heidelber, New York, 1977.
(10.20)	Foith, J. P.	Orientation recognition of randomly oriented workpieces from the shape of their silhouettes Proc. 8th Internat. Symposium on Industrial Robots 1978, p. 584-599. Int. Fluidics Services, Kempston, England.
(10.21)	Foith, J. P.	Picture processing for orientation and position determination for industrial robot applications. IITB Communications 1977, p. 32-40.
(10.22)	Foith, J. P. Geisselmann, H. Lübbert, U. Ringshauser, H.	A modular system for digital imaging sensors for industrial vision. 3rd CISM-IFToMM Symposium on theory and practice of robots and manipulators, Udine, September 1978.
(10.23)	Lübbert, U. Ringshauser, H.	A modular system for television sensors. IITB Communications 1978, p. 9-13.
(10.24)	Gengenbach, O.	Possibilities and practical experience in the application of industrial robots in the metal processing industry. mht 2/75, p. 36-38.
(10.25)	Spur, G. Kraft, H.-R. Siming, H.	Computer-controlled object recognition with optical image sensors. Proceedings 8th Internat. Symposium on industrial robots 1978, Vol. 1, p. 155-164, Int. Fluidics Services, Kempston, England.
(10.26)	Lübbert, U.	Automation of root welding. IITB Communications 1975, p. 64-67.

(10.27)	Bretschi, J. Gudat, H. Lübbert, U. Paul, D. Schief, A.	Industrial sensors for the automation of human sensory faculties in industrial production. FhG reports 1977/1 p. 8-10. Fraunhofer-Gesellschaft, Munich.
(10.28)	Agin, G. I. Duda, R. D.	SRI Vision research for advanced industrial automation. Proceedings of 2nd USA Japan Computer Conference 1975.
(10.29)	Thissen, F.	A device for the automatic optical control of conductor connection patterns for integrated circuits. Philips Technische Rundschau 37, 1977/78, No. 4 p. 85-96.
(10.30)	Baird, M. L.	SIGHT-I: A computer vision system for automated IC chip manufacture. IEEE Trans, systems, man and cybernetics Vol. 8, Feb. 1978, p. 133-139.
(10.31)	Yoda, H. et al.	Direction coding method and its application to scene analysis. Proc. 4th Int. Joint. Conf. on Artificial Intelligence, 1975, p. 620-627.
(10.32)	Takeyasu, K.	An approach to the integrated intelligent robot with multiple sensory feedback: construction and control functions. Proceedings 7th Internat. Symposium on Industrial Robots, Tokyo, p. 523-530.
(10.33)	Kashioka, S.	An approach to the integrated intelligent robot with multiple sensory feedback: visual recognition techniques. Proceedings 7th Internat. Symposium on Industrial Robots, Tokyo, 1977, p. 531-538.
(10.34)	Schulz, D.	Linear and non-linear processing of amplitude-modulated stochastic processes for acoustic quality control and early warning of faults. Thesis University of Karlsruhe, 1976.
(10.35)	Schulz, D.	Acoustic testing of gears. IITB Communications 1976, IITB Karlsruhe, p. 2-5.
(10.36)	No author	Electroacoustic steel converter surveillance. News from ROHDE & SCHWARZ 19, Jan. 1966, p. 21-23.
(10.37)	Maue, H.	An electroacoustic system for process monitoring in oxygen converters. Siemens AG communications, October 1971.
(10.38)	Meier, H. E.	Acoustic quality control of electric motors. IITB Communications 1976, IITB Karlsruhe, p. 6-9.
(10.39)	Roth-Urban, P.	Selection and measurement of operating conditions for sound measurement at hammers and presses. Fortschritte der Akustik p. 274-277, VDI Verlag Dusseldorf 1973.
(10.40)	Roth-Urban, P.	The application of a real-time analyser for measuring sound at machines with non-stationary operating processes messen + prüfen/automatik, Feb. 1974, p. 101-104.

(10.41)	Wang, S. Will, P. M.	Sensors for computer-controlled mechanical assembly. The industrial robot, March 1978, p. 9-18.
(10.42)	Wauer, G.	Positive gripping with an industrial robot gripper with adjustable jaw geometry. Proceedings 8th Internat. Symp. on Industrial Robots 1978, Vol. 1, p. 452-467. Int. Fluidics Services, Kempston, England.
(10.43)	Stojilkovic, Z. Saletic, D.	Learning to recognise patterns by Belgrade hand prothesis. Proceedings 5th Internat. Symposium on Industrial Robots 1975, p. 407-413.
(10.44)	Author's collective	A new handling system as an industrial aid for the working process. 2nd half-yearly report 1976. Item 23, p. 18. Handling systems working group.
(10.45)	Hirt, D. Isenberg, G. Krovinovic, Z.	Development of an adaptive gripping system. Förden und heben 26 (1976) No. 13, Specialist section mht, p. 5-6.
(10.46)	Schweizer, M.	Sensors for program-controlled handling systems. f + h – fördern und heben 27 (1077) No. 4. Specialist section mht, p. 22-27.
(10.47)	Goto, T. et al.	Precise insert operation by tactile controlled robot "HI-T-Hand Expert-2". Proceedings 4th Internat. Symposium on Industrial Robots 1974, Tokyo, p. 209-218.
(10.48)	Fricke, H. W.	Measurement with wire strain gauges. Reprint from Messen + Prüfen, Issue 11/12 Holemann Verlag, Bad Wörishofen.
(10.49)	Schweizer, M. Abele, E.	Sensor-controlled industrial robots for machining tasks. Proceedings 8th Internat. Symposium on Industrial Robots 1978, Stuttgart, p. 904-919.
(10.50)	Leonhards, F.	Measurement transducers for cutting forces and torques, increasing production and overload protection by process control during lathe work. IWF – Report No. 5, p. 187.
(10.51)	Rohrbach, Ch.	Handbook for the electrical measurement of mechanical variables. VDI Verlag Dusseldorf, 1967.
(10.52)	Salmon, M.	The SIGMA system for automation of and machining operations. Montage- und handhabungstechnik 2 (1976), No. 3, 109-112.
(10.53)	Salmon, M.	Assembly by robots. The Industrial Robot, June 1977, p. 81-85.

Chapter 11

| (11.1) | Colding, B.
Colwell, L. V.
Smith, D. N. | Delphi forecasts of manufacturing technology. IFS Publications Ltd. Jan. 1979. |

02/11/